THE GRE**A**

CLIMATE

ROBBERY

How the
food system
drives climate
change and
what we can
do about it

"This book is a must read for movements addressing climate change as well as seed and food Sovereignty. It shows that industrial corporate agriculture is a major part of the climate crisis, and small-scale ecological farming is a significant solution. It also alerts us to the false solutions being offered by those who created the problem — the Exxons of agriculture."
Dr Vandana Shiva, author of *Soil, Not Oil* and *Who Really Feeds the World?*

"Food, land and seeds: protecting them is as essential to climate justice as rooftop solar, wind co-ops, or democratic public transit. This book lifts up the voices of indigenous and peasant farmers around the world, comprehensively explaining why their fight to stop the industrial food juggernaut is the same as the fight for a habitable, just planet."
Naomi Klein, author of *This Changes Everything* and *The Shock Doctrine*

"It's about time that the role agriculture plays in the climate crisis — and the role it could play in the solution — got a concentrated dose of attention. This is fine work that will provoke much new activism!"
Bill McKibben, author of *Deep Economy*

"We welcome the efforts of our colleagues at GRAIN to put Via Campesina's proposals to cool the planet and fight false solutions at the center of the debate. The time has come to change the system, not the climate. Our farmers and indigenous peoples can cool the planet!"
Edgardo García, International Co-ordinating Committee - Via Campesina.

THE GREAT CLIMATE ROBBERY

How the food system drives climate change and what we can do about it

New Internationalist

THE GREAT CLIMATE ROBBERY

Published in 2016 by
New Internationalist Publications Ltd
The Old Music Hall
106-108 Cowley Road
Oxford
OX4 1JE, UK
newint.org

The material in this book was created by GRAIN, which is a small international non-profit organization that works to support small farmers and social movements in their struggles for community-controlled and biodiversity-based food systems. For more information, visit **grain.org**

Cover picture by Pawel Kuczynski (pawelkuczynski.com)

Project Management: Firoze Manji

Copy editing: Andrea Meeson and New Internationalist

Design: Raúl Fernández and New Internationalist

Printed and bound in Britain by TJ International, Cornwall, who hold environmental accreditation ISO 14001.

British Library Cataloguing-in-Publication Data.
A catalogue record for this book is available from the British Library.
Library of Congress Cataloging-in-Publication Data
A catalog record for this book is available from the Library of Congress.

ISBN 978-1-78026-332-8
(ebook ISBN 978-1-78026-340-3)

Contents

Introduction

Governments tend to look at problems through a very narrow lens. In Paris, during the COP21 climate negotiations in December 2015, they said that agriculture is responsible for about 24 per cent of climate change. According to our data, they are missing the bigger picture: the broader food system is actually responsible for around half of all global greenhouse gas emissions, as you will read in this book. Not surprisingly, the solutions they propose to address the problem are also narrowly construed technofixes: from carbon markets and redesigned animal feeds to new genetically modified seeds and top-down forms of agro-ecology. By failing to take a broad systems view and by failing to address the deep political and social inequities underpinning the global food economy, this approach will only get us deeper into trouble.

As we realized on the road to Paris, and even more so during those two weeks of intense mobilization there, it has become ever more critical for people to take matters into their own hands. In addition to challenging the fossil-fuel industry, we need to attack the industrial food system if we want to have a real impact on climate change. This has already begun. In Paris, during the COP21, peasant and indigenous farmers from La Vía Campesina led a symbolic direct action against Danone, one of the world's biggest food and water companies. We took over the public space in front of their headquarters in Paris and painted a red line symbolizing the boundaries that cannot be crossed if we are to have a healthy and just food system – Danone has already crossed these boundaries and needs to be held to account for this. We also held training sessions and workshops engaging thousands of people to talk about how the industrial food system is at the heart of climate change and how people-led food sovereignty, based on local markets and peasant agroecology, can turn all of this around.

This book is meant to support this movement and to help it grow. Over the past 25 years, GRAIN has worked with social movements and organizations around the world to defend local food systems and cultures from the advance of industrial agriculture. Part of our work has involved documenting the ill-effects of this industrial food system – the growing hunger, the destruction of rural people's livelihoods, the loss of biodiversity and cultures, the exploitation of labor and a range of health calamities – and analyzing the ways through which this system expands, from seed laws to free trade agreements to secretive land deals.

But another important part of our work has involved connecting this analysis of the food system to larger issues affecting the planet and linking people's struggles situated within the food system to those happening in other areas. Climate change is one important example of this.

During the past five years, we have pulled together the available data to show how the industrial food system is a major driver of climate change and how food sovereignty is critical to any lasting and just solution. With governments, particularly those from the main polluting countries, abdicating their responsibility to deal with the problem, it has become ever more critical for people to take action into their own hands. Changing the food system is perhaps the most important and effective place to start.

The various articles on climate change selected for this book should provide readers with solid information about how the industrial food system causes climate change, how food and agribusiness corporations are getting away with it and what can be done to turn things around. Other chapters provide a picture of how this climate-killing food system is expanding through the consolidation of corporate control over lands, seeds and markets, and how struggles are under way to stop it.

We hope this book will help readers to better understand the ways in which corporations seek to increase their control over the food system so that this control can be more effectively challenged. We hope it will inspire people to take action and that it will provide readers with some information and analysis that they can use directly in their own work.

When the publishers asked us to provide the name of the book's editor in order to help it gain wider distribution, we reluctantly agreed. The reluctance arose from the fact that this book is the result of the collective effort of all GRAIN's staff, many of our partners across the world, and the numerous translators, editors and volunteers who help us in our work. Without all these people, this book would not have seen the light.

Henk Hobbelink, for GRAIN

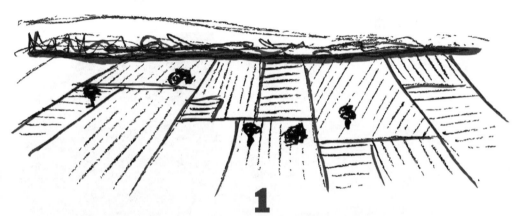

1
Food and climate change:
the forgotten link

1.1 How the industrial food system

Between 44% and 57% of all greenhouse gas (GHG)

Deforestation: 15%-18%

Before the planting starts, the bulldozers do their job. Worldwide, industrial agriculture is pushing into savannas, wetlands and forests, plowing under huge amounts of land. The FAO says the expansion of the agricultural frontier accounts for between 70 per cent and 90 per cent of global deforestation, at least half of that for the production of a few agricultural commodities for export. Agriculture's contribution to deforestation thus accounts for between 15 per cent and 18 per cent of global GHG emissions.

Other non-food related emissions 43%-56%

Waste: 3%-4%

The industrial food system discards up to half of all the food that it produces, thrown out on the long journey from farms to traders, to food processors, and eventually to retailers and restaurants. A lot of this waste rots on garbage heaps and landfills, producing substantial amounts of GHGs. Between 3.5 per cent and 4.5 per cent of global GHG emissions come from waste and more than 90 per cent of these are produced by materials originating within the food system.

Freezing and Retail: 2%-4%

Refrigeration is the lynchpin of the modern supermarket and fast food chains' vast global procurement systems. Wherever the industrial food system goes, so do cold chains. Considering that cooling is responsible for 15 per cent of all electricity consumption worldwide, and that leaks of chemical refrigerants are a major source of GHGs, we can safely say that the refrigeration of foods accounts for about one per cent to two per cent of all global greenhouse gas emissions. The retailing of foods accounts for about the same.

contributes to the climate crisis

emissions come from the global food system

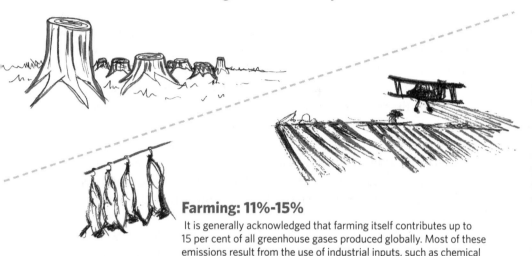

Farming: 11%-15%

It is generally acknowledged that farming itself contributes up to 15 per cent of all greenhouse gases produced globally. Most of these emissions result from the use of industrial inputs, such as chemical fertilizers and petrol to run tractors and irrigation machinery, as well as the excess manure generated by intensive livestock keeping.

Transport: 5%-6%

The industrial food system acts like a global travel agency. Much of our food, grown under industrial conditions in faraway places, travels thousands of kilometers before it reaches our plates. Crops for animal feed may be grown in Argentina and fed to chickens in Chile that are exported to China for processing and eventually eaten in a McDonalds in the US. We can conservatively estimate that the transportation of food accounts for one quarter of global GHG emissions linked to transportation, or between five and six per cent of all global GHG emissions.

Processing and packaging: 8%-10%

Processing is the next, highly profitable step in the industrial food chain. The transformation of foods into ready-made meals, snacks and beverages requires an enormous amount of energy, mostly in the form of carbon. So does the packaging and canning of these foods. Processing and packaging enables the food industry to stack the shelves of supermarkets and convenience stores with hundreds of different formats and brands, but it also generates a huge amount of greenhouse gas emissions — some eight per cent to 10 per cent of the global total.

Food and climate: piecing the puzzle together

Agriculture is starting to get more attention in the international negotiations around climate change. The consensus is that it contributes about 10 per cent to 15 per cent of all global human-made greenhouse gas emissions (GHGs), making it one of the key drivers of climate change. But there's no point in talking about agriculture without looking at the larger food system.

Beyond the emissions that occur on the farm, today's dominant industrial food system generates GHGs by transporting food around the world, by deforesting to make way for plantations, and by generating waste. Pulling together the available data on these sources of emissions reveals that the global food system is responsible for about half of all global GHGs. So, when it comes to climate change, the food system is at the center of the problem.

Most studies put the contribution of agricultural emissions – the emissions produced on the farm – at somewhere between 11 per cent and 15 per cent of all global emissions.[1,2] What often goes unsaid, however, is that most of these emissions are generated by industrial farming practices that rely on chemical (nitrogen) fertilizers, heavy machinery run on petrol, and highly concentrated industrial livestock operations that pump out methane.

The figures for agriculture's contribution also often do not account for its role in land-use changes and deforestation, which are responsible for nearly one-fifth of global GHG emissions.[3,4] Worldwide, agriculture is pushing into savannas, wetlands, cerrados and forests, and plowing under huge amounts of land. The expansion of the agricultural frontier is the dominant contributor to deforestation, accounting for between 70 per cent and 90 per cent of global deforestation.[5,6] This means that up to 18 per cent of global GHG emissions are produced by land-use change and deforestation caused by agriculture. And here too, the global food system and the industrial model of agriculture are the chief culprits. The main driver of this deforestation is the expansion of industrial plantations for the production of commodities, such as soy, sugarcane, oil palm, maize and rapeseed. Since 1990, the area planted with these five commodity crops grew by a whopping 38 per cent.[7]

These emissions from agriculture account for only a portion of the food system's overall contribution to climate change. Equally important is what happens when food leaves the farm until it enters our bodies.

Food and climate change: the forgotten link

Food is the world's biggest economic sector, involving more transactions and employing more people by far than any other. These days food is prepared and distributed using enormous amounts of processing, packaging and transportation, all of which generate GHG emissions, although data on such emissions are hard to find. Studies looking at the EU conclude that about one quarter of overall transportation involves commercial food transport.[8] The scattered figures on transportation available for other countries, such as Kenya and Zimbabwe, indicate that the percentage is even higher in non-industrialized countries, where "food production and delivery accounts for between 60 per cent and 80 per cent of the total energy – human plus animal plus fuel – used."[9] With transportation accounting for 25 per cent of global GHG emissions, we can use the EU data to conservatively estimate that the transport of food accounts for at least six per cent of global GHG emissions.

When it comes to processing and packaging, again the available data is mainly from the EU, where studies show that the processing and packaging of food accounts for about 10 per cent of GHG emissions,[10] while refrigeration of food accounts for up to four per cent[11] of total emissions and food retail another two per cent.[12,13,14] Playing it conservative with the EU figures and extrapolating from the scarce figures that exist for other countries, we can estimate that at least five per cent of emissions are due to food transport, up to 10 per cent due to food processing and packaging, around two per cent due to refrigeration, and two per cent due to retail. This gives us a total contribution of between 15 and 20 per cent of global emissions from these activities.

Not all of what the food system produces gets consumed. The industrial food system discards up to half of all the food that it produces, in its journey from farms to traders, to food processors, to stores and supermarkets. This is enough to feed the world's hungry six times over.[15] A lot of this waste rots away on garbage heaps and landfills, producing substantial amounts of greenhouse gases. Different studies indicate that somewhere between 3.5 and 4.5 per cent of global GHG emissions come from waste, and that more than 90 per cent of them come from materials originated in agriculture and their processing.[16] This means that the decomposition of organic waste originated in food and agriculture is responsible for three per cent to four per cent of global GHG emissions.

Add all the above factors up, and there is no escape from the conclusion that the current global food system, propelled by an increasingly powerful transnational food industry, is responsible for about half of all human-produced greenhouse gas emissions: anywhere between a low of 44 per cent to a high of 57 per cent.

Sources:

GRAIN 2011 "Food and climate change: the forgotten link": https://www.grain.org/e/4357
GRAIN 2013 "Food, climate change and healthy soils", in *UNCTAD Trade and Environment Review 2013*, Chapter 1, Commentary IV, p.19[17]

1 See 'Climate Change 2007: Mitigation of Climate Change. Chapter 8: Agriculture', IPCC Fourth Assessment Report (AR4), http://tinyurl.com/ms4mzb
2 Legg, W. and Huang, H. (2010) "OECD Trade and Agriculture Directorate, Climate change and agriculture", *OECD Observer*, No. 278, http://tinyurl.com/5u2hf8k
3 *Ibid.*
4 IPCC (2004). Climate Change 2001: Working Group I: 3.4.2 "Consequences of Land-use Change", http://tinyurl.com/6lduxqy
5 See FAO Advisory Committee on Paper and Wood Products – Session 49 (2008), Bakubung, South Africa, 10 June; and
6 Kanninen, M. et al. (2007) "Do trees grow on Money?", *Forest Perspective 4*, CIFOR, Jakarta, cifor.org/publications/pdf_files/Books/BKanninen0701.pdf
7 See GRAIN (2010), "Global Agribusiness: two decades of plunder", in *Seedling*, July, https://grain.org/article/entries/4055-global-agribusiness-two-decades-of-plunder
8 See: Eurostat (2011) "From farm to fork – a statistical journey along the EU's food chain", ec.europa.eu/eurostat/en/web/products-statistics-in-focus/-/KS-SF-11-027
9 Karekezi, S. and Lazarus, M. (1995) *Future energy requirements for Africa's Agriculture*. Chapters 2, 3, and 4, fao.org/docrep/V9766E/v9766e00.htm#Contents
10 For EU, see Bolla, V. and Pendolovska, V. (2011), "Driving forces behind EU-27 greenhouse gas emissions over the decade 1999-2008", http://tinyurl.com/6bhesog
11 Garnett, T. and Jackson, T. (2001) "Bitten: an exploration of refrigeration dependence in the UK food chain and its implications for climate policy", Food Climate Research Network, Centre for Environmental Strategy, University of Surrey, 1 June, fcrn.org.uk/fcrn/publications/frost-bitten
12 Tassou, SA, et al (2011), "Energy consumption and conservation in food retailing", Applied Thermal Engineering 31: 147-56
13 Venkat, K. (2012) "The Climate Change Impact of US Food Waste", *Int.J. Food System Dynamics* 2(4): 431-46, cleanmetrics.com/pages/ClimateChangeImpactofUSFoodWaste.pdf
14 Bakas, I (2010) "Food and Greenhouse Gas (GHG) Emissions", Copenhagen Resource Institute, scp knowledge.eu/sites/default/files/KU_Food_GHG_emissions.pdf
15 Stuart, Tristam (2009) *Waste: Uncovering the Global Food Scandal*, Penguin, http://tinyurl.com/m3dxc9

16 Bogner, J. et al. (2008) "Mitigation of global greenhouse gas emissions from waste: conclusions and strategies from the IPCC Fourth Assessment Report. Working Group III (Mitigation)", wmr.sagepub.com/content/26/1/11.short?rss=1&ssource=mfc

17 See: unctad.org/en/pages/PublicationWebflyer.aspx?publicationid=666

1.2 Food sovereignty: five steps to

1 Take care of the soil

The food/climate equation is rooted in the earth. The expansion of unsustainable agricultural practices over the past century has led to the destruction of between 30 per cent and 75 per cent of the organic matter on arable lands, and 50 per cent of the organic matter on pastures and prairies. This massive loss of organic matter is responsible for between 25 per cent and 40 per cent of the current excess CO_2 in the earth's atmosphere. But the good news is that this CO_2 that we have sent into the atmosphere can be put back into the soil, simply by restoring the practices that small farmers have been engaging in for generations.

If the right policies and incentives were in place worldwide, soil organic matter contents could be restored to pre-industrial agriculture levels within a period of 50 years – which is roughly the same time frame that industrial agriculture took to reduce it. This would offset between 24 per cent and 30 per cent of all current global greenhouse gas emissions.

3 Cut the food miles, and focus on fresh food

The corporate logic that results in the shipment of foods around the world and back again, makes no sense from an environmental perspective, or any other perspective for that matter. The global trade in food, from the opening of vast swaths of lands and forests to produce agricultural commodities to the frozen foods sold in supermarkets, is the chief culprit in the food system's overweight contribution to GHG emissions. Much of the food system's GHG emissions can be eliminated if food production is reoriented towards local markets and fresh foods, and away from cheap meat and processed foods. But achieving this is probably the toughest fight of all, as corporations and governments are deeply committed to expanding the trade in foods.

5 Forget the false solutions, focus on what works

There is growing recognition that food is central to climate change. The latest IPCC reports and international summits have recognized that food and agriculture are major drivers of GHG emissions and that climate change poses tremendous challenges to our capacity to feed a growing global population. Yet there

cool the planet and feed its people

2 Natural farming, no chemicals

The use of chemicals on industrial farms is increasing all the time, as soils are further depleted and pests and weeds become immune to insecticides and herbicides. Small farmers around the world, however, still have the knowledge and the diversity of crops and animals to farm productively without the use of chemicals, by diversifying cropping systems, integrating crop and animal production, and incorporating trees and wild vegetation. These practices enhance the productive potential of the land because they improve soil fertility and prevent soil erosion. Every year more organic matter is built up in the soil, making it possible to produce more and more food.

4 Give the land back to the farmers, and stop the mega plantations

Over the past 50 years, a staggering 140 million hectares – the size of almost all the farmland in India – has been taken over by four crops grown predominantly on large plantations: soybeans, oil palm, rapeseed and sugar cane. The global area under these and other industrial commodity crops, all of them notorious emitters of greenhouse gases, is set to further grow if policies don't change. Today, small farmers are squeezed onto less than one quarter of the world's farmlands, but they continue to produce most of the world's food – 80 per cent of the food in non-industrialized countries, according to the FAO. Small farmers produce this food far more efficiently than big plantations, and in ways that are better for the planet. A worldwide redistribution of lands to small farmers, combined with policies to help them rebuild soil fertility and policies to support local markets, can reduce GHG emissions by half within a few decades.

has been zero political will to challenge the dominant model of industrial food production and distribution. Instead, governments and corporations are proposing a number of false solutions. There is the empty shell of Climate Smart Agriculture, which is essentially just a rebranding of the Green Revolution. There are new, risky technologies, such as crops genetically engineered for drought resistance or large-scale geo-engineering projects. There are mandates for biofuels, which are driving land grabs in the South. And there are carbon markets and REDD+ projects, which essentially allow the worst GHG offenders to avoid cuts in emissions by turning the forests and farmlands of peasants and indigenous peoples into conservation parks and plantations. None of these "solutions" can work because they all work against the only effective solution: a shift from a globalized, industrial food system governed by corporations to local food systems in the hands of small farmers.

The original poster can be downloaded from: https://www.grain.org/e/5102

Turning the food system upside down

If measures are taken to restructure agriculture and the larger food system, based on food sovereignty, small-scale farming, agro-ecology and local markets, we can cut global emissions in half within a few decades. We don't need carbon markets or techno-fixes. We need the right policies and programs to dump the current industrial food system and create a sustainable, equitable and truly productive one instead. Clearly, we will not get out of the climate crisis if the global food system is not urgently and dramatically transformed. The place to start is with the soil.

Food begins and ends with soil. It grows out of the soil and eventually goes back in it to enable more food to be produced. This is the very cycle of life. But in recent years humans have ignored this vital cycle. We have been taking from the soil without giving back. The industrialization of agriculture, which started in Europe and North America and was later replicated in the Green Revolution that took place in other parts of the world, was based on the assumption that soil fertility can be maintained and increased through the use of chemical fertilizers. Little attention was paid to the importance of organic matter in the soil.

A wide range of scientific reports indicate that cultivated soils have lost between 30 per cent and 75 per cent of their organic matter during the 20th century, while soils under pastures and prairies have typically lost up to 50 per cent. There is no doubt that these losses have provoked a serious deterioration of soil fertility and productivity, as well as worsening droughts and floods. Taking as a basis some of the most conservative figures provided by scientific literature, the global accumulated loss of soil organic matter over the last century can be estimated between 150 to 200 billion tonnes.[1] Not all this organic matter ended up in the air as CO_2, as significant amounts have been washed away by erosion and have been deposited in the bottom of rivers and oceans. However, it can be estimated that at least 200 to 300 billion tonnes of CO_2 have been released to the atmosphere due to the global destruction of soil organic matter. In other words, 25 per cent to 40 per cent of the current excess of CO_2 in the atmosphere comes from the destruction of soils and its organic matter.

There is some good news hidden in these devastating figures. The CO_2 that we have sent into the atmosphere by depleting the world's soils can be put back into the soil. All that is required is a change of agricultural practices.

We have to move away from practices that destroy organic matter to practices that build up the organic matter in the soil. We know this can be done. Farmers around the world have been engaging in these very practices for generations. GRAIN research has shown that if the right policies and incentives were in place worldwide, soil organic matter contents could be restored to pre-industrial agriculture levels within a period of 50 years – which is roughly the same time frame that industrial agriculture took to reduce it.[2] The continuing use of these practices would allow the offset of between 24 per cent and 30 per cent of current global annual GHG emissions.[3]

The new scenario would require a radical change in approach from the current industrial agriculture model. It would focus on the use of techniques such as diversified cropping systems, better integration between crop and animal production, increased incorporation of trees and wild vegetation, and so on. Such an increase in diversity would, in turn, increase the production potential and the incorporation of organic matter would progressively improve soil fertility, creating virtuous cycles of higher productivity and higher availability of organic matter. The capacity of soil to hold water would increase, which would mean that excessive rainfall would lead to fewer, less intense floods and droughts. Soil erosion would become less of a problem. Soil acidity and alkalinity would fall progressively, reducing or eliminating the toxicity that has become a major problem in tropical and arid soils. Additionally, increased soil biological activity would protect plants against pests and diseases. Each one of these effects implies higher productivity and hence more organic matter available to soils, thus making possible, as the years go by, higher targets for soil organic matter incorporation. More food would be produced in the process.

To be able to do it, we would need to massively build on the skills and experience of the world's small farmers, rather than undermining them and forcing them off their lands, as is now the case. A global shift towards an agriculture that builds up organic matter in the soil would also put us on a path to resolving some of the other major sources of GHGs from the food system. Basically there are three other mutually reinforcing shifts that need to take place in the food system to address its overall contribution to climate change. The first is a shift to local markets and short-circuits of food distribution, which will cut back on transportation and the need for packaging, processing and refrigeration. The second is a reintegration of crop and animal production, to cut back on transportation, the use of chemical fertilizers and the production of methane and nitrous oxide emissions generated by intensive meat and dairy operations. And the third is the stopping of land clearing and deforestation, which will require genuine agrarian reform and a reversal of the expansion of monoculture plantations for the production of agrofuels and animal feed.

If the world gets serious about putting these four shifts into action, it is quite possible that we can cut global GHG emissions in half within a few decades and, in the process, go a long way towards resolving the other crises affecting the planet, such as poverty and hunger. There are no technical hurdles standing in the way – the knowledge and skills are in the hands of the world's farmers and we can build on that. The only hurdles are political, and this is where we need to focus our efforts.

Sources:

GRAIN 2011 "Food and climate change: the forgotten link":
https://www.grain.org/e/4357
GRAIN 2013 "Food, climate change and healthy soils", in *UNCTAD Trade and Environment Review 2013*, Chapter 1, Commentary IV, p.19[4]

1 Figures used for calculations were: a) an average loss of 4.5-6 kg of SOM/m^2 of arable land and 2-3kg of SOM/m^2 of agricultural land under prairies and not cultivated; b) an average soil depth of 30cm, with an average soil density of $1 gr/cm^3$; c) 5,000 million ha of agricultural land worldwide; 1,800 million ha of arable land, as stated by FAO; and d) a ratio of 1.46kg of CO_2 for each kilogram of destroyed SOM
2 See: GRAIN (2009) "Earth matters: tackling the climate crisis from the ground up", in *Seedling*, October, grain.org/e/735
3 The conclusion is based on the assumption that organic matter incorporation would reach an annual global average rate of 3.5 to 5 tonnes per hectare of agricultural land. For more detailed calculations, see: Table 2 in GRAIN (2009), *Ibid.*
4 See: http://unctad.org/en/pages/PublicationWebflyer.aspx?publicationid=666

1.3 The Exxons of agriculture

It goes without saying that oil and coal companies should not have a seat at the policy table for decisions on climate change. Their profits depend on business-as-usual and they'll do everything in their power to undermine meaningful action.

But what about fertilizer companies? They are essentially the oil companies of the food world: the products they get farmers to pump into the soil are the largest source of emissions from farming.[1] They, too, have their fortunes wrapped in agribusiness-as-usual and the expanded development of cheap sources of energy, such as shale gas.

Exxon and BP must envy the ease their fertilizer counterparts have had in infiltrating the climate change policy arena. In the governmental climate negotiations there is only one major intergovernmental initiative that has emerged to deal with climate change and agriculture – and it is controlled by the world's largest fertilizer companies.

The Global Alliance for Climate Smart Agriculture, launched in 2014 at the United Nations (UN) Summit on Climate Change in New York, is the culmination of several years of efforts by the fertilizer lobby to block meaningful action on agriculture and climate change. Of the Alliance's 29 non-governmental founding members, there are three fertilizer industry lobby groups, two of the world's largest fertilizer companies (Yara of Norway and Mosaic of the US), and a handful of organizations working directly with fertilizer companies on climate change programs. Today, 60 per cent of the private sector members of the Alliance still come from the fertilizer industry.[2]

Corporate smart agriculture

One possible explanation for the fertilizer industry's successful policy coup is that its role in climate change is poorly understood and severely underestimated. People associate Shell, not Yara, with fracking. But it is Yara that co-ordinates the corporate lobby for shale gas development in Europe, and it is Yara and other fertilizer companies that suck up most of the natural gas produced by the fracking boom in the US.[3]

Fertilizers, especially nitrogen fertilizers, require an enormous amount of energy to produce. Estimates are that fertilizer production accounts for between one and two per cent of total global energy consumption and produces about the same share of global greenhouse gas (GHG) emissions.[4] This production gets bigger every year. Supplies of nitrogen fertilizer, which is produced almost entirely from natural gas, are expected to grow nearly four per cent per year over the next decade.[5] And this production will increasingly rely on natural gas from fracked wells, which leak 40 per cent to 60 per cent more methane than conventional natural gas wells. (Methane is 25 times more potent than CO_2 as a greenhouse gas.)[6]

Production, however, accounts for only a small fraction of the GHG emissions generated by chemical fertilizers. Most emissions occur once they are applied to the soil.

The International Panel on Climate Change (IPCC) estimates that for every 100kg of nitrogen fertilizer applied to the soil, one kilogram ends up in the atmosphere as nitrous oxide (N_2O), a gas that is 300 times more potent than CO_2 as a greenhouse gas and is the world's most significant ozone-depleting substance. In 2014, this was equivalent to the average annual emissions of 72 million cars driven in the US – about one third of the US fleet of cars and trucks.[7]

New research, however, shows that these alarming numbers are at least three to five times too low. The use of chemical fertilizers this year will likely generate more GHG emissions than the total emissions from all of the cars and trucks driven in the US (*See Box 1: The fertilizer footprint on page 19*).

The fertilizer industry has long known that its chemicals are cooking the planet and there is a growing body of evidence that shows that its products are not needed to feed the world. Farmers can stop using chemical fertilizers without reducing yields by adopting agro-ecological practices.[8] This was the conclusion supported by the 2008 International Assessment of Agricultural Knowledge, Science and Technology for Development (IAASTD) – a three-year intergovernmental process involving more than 400 scientists and sponsored by the World Bank and all of the relevant UN agencies.[9]

Faced with this dilemma, the fertilizer companies have moved aggressively to control the international debate on agriculture and climate change, and to position themselves as a necessary part of the solution.

Fronting for fertilizers

There have been several organizations advocating at the international level for sustainable agriculture to be interpreted as synonymous with agro-ecology. However, agro-ecology has unfortunately come to represent principles which

14

reject the use of farming inputs. Therefore, initiatives such as the Global Alliance for Climate Smart Agriculture are important to ensure the UN system adopts decisions that are reflective of modern agriculture.

Canadian Federation of Agriculture[10]

The global fertilizer industry is dominated by a handful of corporations. Yara, which is more than 40-per-cent owned by the Norwegian government and its state pension fund, dominates the global market for nitrogen fertilizer, while US-based Mosaic and a few companies in Canada, Israel and the former Soviet Union operate cartels that control the global potash supply. Mosaic is also the leading producer of phosphates.

These companies are collectively represented by a number of lobby groups. The main ones at the global level are The Fertilizer Institute, the International Fertilizer Industry Association and the International Plant Nutrition Institute. Fertilizer companies are also represented by energy consumer lobby groups, such as the International Federation of Industrial Energy Consumers. Yara chairs the latter's Gas Working Party, which, in collaboration with Fertilizers Europe, is lobbying heavily for shale gas development in the European Union (EU).[11]

Graph 1 World's 10 largest fertilizer companies

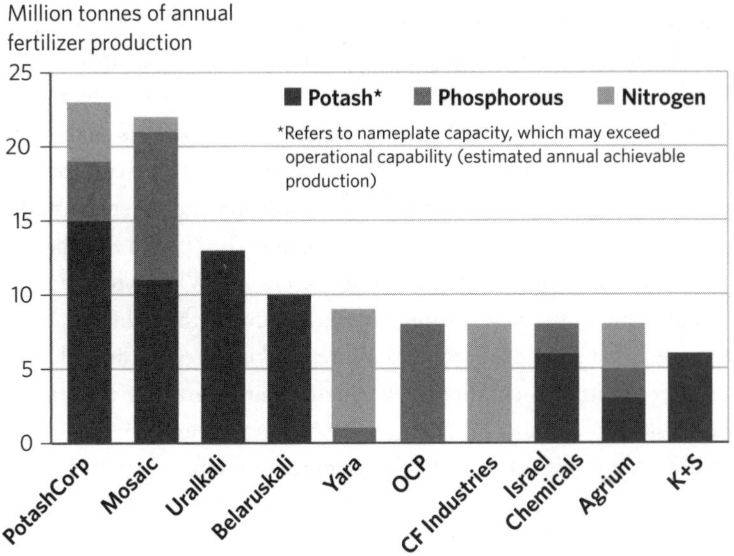

Source: Fertecon, CRU, Company Reports, PotashCorp

The fertilizer companies and their front groups play an active role in various alliances that they have formed with other corporations from the food and agriculture sectors to define and protect their collective interests on policies related to the environment and climate change.[12]

In North America, for instance, Yara and other fertilizer companies and lobby groups co-founded the Alliance for Sustainable Agriculture (Field To Market) alongside other major food and agribusiness companies like Walmart, Kelloggs and Monsanto. Also active in this alliance are big US environmental non-governmental organizations (NGOs), such as the Environmental Defense Fund (EDF) and the The Nature Conservancy (TNC). These NGOs work directly with Yara, Mosaic and other fertilizer companies on "climate smart" fertilizer efficiency programs that Walmart, PepsiCo, Campbells and other major food companies and retailers are using as a basis for their internal GHG emissions reduction plans (*See Box 2: Pollution as the solution on page 21*).

The same NGOs and fertilizer front groups are behind Solutions from the Land, a US alliance of agribusiness corporations and corporate farmers established to defend industrial agriculture from environmental regulations, initially dealing with the destructive impacts on waterways from chemical fertilizer run-off and now focusing on climate change.

"We're scared to death we'll get hijacked by some groups that oppose technology," explains Solution from the Land's Fred Yoder, speaking in Abu Dhabi in March 2015 at an agribusiness forum on climate change.[13]

In early 2015, Solutions from the Land changed its name to the North American Alliance for Climate Smart Agriculture and now acts as the regional co-ordination for the Global Alliance for Climate Smart Agriculture.

This cozy relationship between the fertilizer industry and other transnationals of the food and agribusiness sector reaches beyond the US and Europe. Yara is particularly active within the World Economic Forum (WEF) where it co-chairs the development of its New Vision for Agriculture with Walmart. Yara also chairs the WEF's Climate Smart Agriculture working group, through which it co-ordinates the implementation of "climate smart" fertilizer programs with Nestlé, PepsiCo, Syngenta and other companies in Asia and Africa.

Fertilizer companies also have a long-standing relationship with the international research centers of the Consultative Group for International Agricultural Research (CGIAR). Today, the fertilizer industry collaborates with these centers on various climate smart initiatives in the South (*See Box 2:*

Pollution as the solution on page 21). The relationship extends to the Bill Gates-funded Alliance for a Green Revolution in Africa (AGRA), which has several areas of co-operation with the CGIAR and the fertilizer industry, such as the African Green Revolution Forum that was established by Yara and AGRA in 2010.

The main vehicle for the promotion of fertilizers in the South, however, is the International Fertilizer Development Center (IFDC), which was established in the US in the 1970s and is funded by several fertilizer companies, including Yara. The IFDC lobbies governments for policies that increase fertilizer use and promote different fertilizer application techniques, such as integrated soil management that AGRA, the World Bank and other funding agencies have embraced as "climate smart".

All of these various corporations, agencies, front groups and alliances have converged behind a common effort to promote "climate smart agriculture" as the official response to climate change. It builds upon previous, equally abstract terms promoted by the fertilizer industry to cast chemical fertilizers as part of the solution to climate change, such as "climate compatible agricultural growth" and "sustainable intensification".[14]

"I believe 2015 and 2016 will be the years where we move from building a global movement to action on the ground. And the key words are climate smart agriculture, an area where Yara has products and knowledge," says Sean de Cleene, Vice President, Global Initiatives, Strategy and Business Development at Yara.[15]

The UN's Food and Agriculture Organization (FAO) first coined the term "climate smart agriculture" in 2010 as a means to attract climate finance to its agricultural programs in Africa. The term only became significant in international policy circles in 2012 after the second Global Conference on Agriculture, Food Security and Climate Change, organized in Hanoi by the World Bank and FAO and hosted by the government of Vietnam.

The choice of Vietnam was no accident. Yara and other food and agribusiness transnationals of the WEF had recently launched a major public-private partnership with the Vietnamese government under which these corporations were given exclusive responsibility over the "value chains" of the country's main export commodities. Yara was put in charge of coffee and vegetables, and the programs in Vietnam were adopted as the WEF's first pilot project for climate smart agriculture, which Yara was tasked with overseeing.[16]

The program of the Second Global Conference was dominated by Yara and the other corporations collaborating with the Vietnamese government. Civil society organizations were marginalized from the discussions, and their vocal rejection of the "climate smart agriculture" concept was ignored.[17] While the previous conference had called for a "paradigm shift at all levels", this time the conference ended with a call for "a paradigm shift in the role of the private sector" to "institutionalize and scale up" private sector involvement and "move from public-private to private-public partnerships".[18]

By the time of the next Global Conference in South Africa a year later, the fertilizer lobby and its allies had produced a plan for the creation of an Alliance for Climate Smart Agriculture to be formally presented at the UN Climate Summit in September 2014 as the international community's main platform for action on climate change and agriculture.

The US State Department then took the lead in moving the plan forward. At the Alliance's "Partner Meeting" in The Hague in July 2014, where the final details were hammered out, the US sent five government officials, four representatives of US agribusiness lobby groups and four corporate representatives – a number equal to the entire number of delegates from developing countries.[19]

"The international discussions were hijacked by agribusiness companies, the World Bank, the US and other climate smart agriculture-friendly governments," says World Food Prize winner Hans Herren. "They have the money and the lobby groups. Those of us defending agro-ecology, local food systems and small-scale farming as the holistic and truly climate-friendly solution were simply pushed out of the process."[20]

Today the Global Alliance for Climate Smart Agriculture is stacked with fertilizer companies, fertilizer front groups and NGOs and companies that work directly with them (see Graph 2, page 23). Its steering committee includes Yara, Mosaic, EDF and TNC, as well as their home governments of Norway and the US.[21]

Back to a paradigm shift

Food and agriculture are low-hanging fruits for action on climate change. Dramatic and rapid reductions in GHG emissions can be achieved in our food systems without major economic consequences. The elimination of chemical fertilizers is one of the easiest and most effective places to start.

Cutting out chemical fertilizers could reduce annual global greenhouse emissions by as much as 10 per cent (*See Box 1: The fertilizer footprint below*). Additionally, the shift from chemical fertilizers to agro-ecological practices would allow farmers to rebuild organic matter in the world's soils, and thus capture a possible two-thirds of the current excess CO_2 in the atmosphere within 50 years.[22] There are also the added benefits of improved livelihoods for farmers, more nutritious foods, protection of the ozone layer and safe water systems.

No technical hurdles stand in the way. Fertilizer companies may claim that if we stopped using their products we would have to plow up the earth's remaining forests in order to meet global food needs, but there are plenty of studies showing that farmers using simple agro-ecological practices can produce as much food without chemical fertilizers on the same amount of land.

When it comes to global food security, we should be much more worried about our dependence on the cartels that the fertilizer companies operate. During the 2007 food price crisis, as one billion people starved because they could no longer afford food, the fertilizer companies jacked up their prices and held governments and farmers at ransom. They pointed to rising costs for raw materials (natural gas) but the profits of Yara and Mosaic jumped a staggering 100 per cent that year.[23]

Kicking the fertilizer habit is really a matter of politics. No meaningful action can occur until the fertilizer industry's grip on policymakers is loosened. Let's start making this happen by shutting down the Global Alliance for Climate Smart Agriculture.

Box 1: The fertilizer footprint

Scientists now know that the 17% increase of N_2O in the atmosphere since the pre-industrial era is a direct result of chemical fertilizers, owing especially to the deployment of the so-called Green Revolution programs of the 1960s that brought chemical fertilizers into use in Asia and Latin America.[24] They also now know that the amount of N_2O emissions resulting from the application of nitrogen fertilizers is more in the range of three per cent to five per cent, a dramatic increase from the IPCC's assumption of one per cent.[25]

Yet even this estimate does not go far enough in assessing current and future emissions from fertilizers. First, fertilizer use is expanding fastest in the tropics, where soils generate even higher rates of N_2O emissions per kilogram of nitrogen applied, particularly when the soils have been deforested.[26] Secondly, fertilizer use per hectare is growing and new studies show that the rate of N_2O emissions increases exponentially as more fertilizer is applied.[27]

Chemical fertilizers are addictive. Because they destroy the natural nitrogen in the soils that is available to plants, farmers have to use more and more fertilizers every year to sustain yields. Over the past 40 years, the efficiency of nitrogen fertilizers has decreased by two-thirds and their consumption per hectare has increased by seven times.[28,29]

The effect on organic matter, the world's most important carbon sink, is the same. Despite industry propaganda to the contrary, recent studies demonstrate that chemical fertilizers are responsible for much of the massive loss of organic matter that has occurred in the world's soils since the pre-industrial era.[30]

"In numerous publications spanning more than 100 years and a wide variety of cropping and tillage practices, we found consistent evidence of an organic carbon decline for fertilized soils throughout the world," says University of Illinois soils scientist Charlie Boast.[31]

Soils around the world have lost, on average, at least one to two percentage points of organic matter in the top 30cm since chemical fertilizers began to be used. This amounts to some 150,000–205,000 million tonnes of organic matter, which has resulted in 220,000–330,000 million tonnes of CO_2 emitted into the air or 30 per cent of the current excess CO_2 in the atmosphere! [32]

The overall contribution of chemical fertilizers to climate change has thus been drastically underestimated and a reassessment is urgently needed. Factoring in the recent research, the growing reliance on shale gas and the impacts on soil organic matter could push estimates of the share of global GHG emissions from chemical fertilizers to as high as 10 per cent. The world needs to move quickly to end our deadly addiction to these toxic products.

Box 2: Pollution as the solution

There is no precise definition for climate smart agriculture, and deliberately so. The Global Alliance for Climate Smart Agriculture instead leaves it to its members to determine what climate smart agriculture means to them.[33]

"Membership in the Alliance does not create any binding obligations and each member individually determines the nature of its participation," states the Alliance's brochure.[34] So what are these climate smart agriculture programs that the Alliance members are pursuing?

The FAO, one of the leading organizers of the Alliance, produced a sourcebook and an accompanying list of 10 climate smart agriculture "success stories". All of the examples are top-down extension programs, including a nitrogen fertilizer application technique promoted by the IFDC, which focuses on small-scale farmers in the South whose contributions to climate change are negligible.[35]

The CGIAR has a similar set of climate smart "success stories" that focus on the South, promote the use of fertilizers and genetically modified organisms, and make no mention of agro-ecology.[36] Some of the CGIAR centers are already working directly with the fertilizer industry and other agribusiness companies on climate smart projects. The International Maize and Wheat Improvement Center, for example, has a Climate-Smart Villages project with the fertilizer industry's International Plant Nutrition Institute to help farmers in Africa and Asia "identify fertilizer options".[37]

Most climate smart agriculture initiatives, however, come directly from the private sector, through alliances between the major agribusiness and food companies. The US government, which says its "climate smart agriculture" strategy will be "voluntary and incentive based", cites 10 cases of private sector initiatives in line with its strategy. Three of these programs are based on "fertilizer optimization": one called Field to Market through the Alliance for Sustainable Agriculture (a network of the largest food and agribusiness companies), a second called 4R that is run by The Fertilizer Institute and The Nature Conservancy, and a third that is a collaboration between Walmart, the Environmental Defense Fund and one of the biggest fertilizer distributors in the US.[38]

Walmart's climate smart agriculture program is particularly significant, since it is the world's biggest food retailer. Walmart intends to achieve most of its targeted GHG emission reductions by enrolling its suppliers in "fertilizer optimization" programs developed by Yara and other fertilizer companies and their NGO partners. So far, Walmart has secured commitments from Campbell Soup, Cargill, Dairy Farmers of America, General Mills, Monsanto, Kelloggs, PepsiCo, Smithfield Foods and Unilever to implement these programs in their supply chains.[39,40,41]

What this means on the ground can be seen in the model project that Yara is implementing with PepsiCo on the plantations that supply oranges for its Tropicana juices. Under the project, PepsiCo gets these plantations to purchase Yara's "low carbon footprint" branded nitrogen fertilizers, which are supposed to produce less fertilizer run-off. These "premium branded fertilizers" were developed by Yara "in order to avoid a situation where only organically produced food would gain the climate brand of approval".[42,43]

In Africa, where much of the attention of the Global Alliance is perversely focused, the fertilizer industry and its allies maintain that increasing the use of fertilizers is a climate smart way to reduce greenhouse gas emissions. Yara and Syngenta are running trials in Tanzania to show that increasing yields with chemical fertilizers and hybrid seeds "reduces the need for deforestation, thereby avoiding GHG emissions".[44] This is what they refer to as "sustainable intensification", a concept that the FAO categorizes as climate smart.

Africa is not merely of interest to the fertilizer industry as a way to deflect attention from agricultural emissions in the North. It is the world's fastest-growing market for chemical fertilizers and an important new source of natural gas reserves, especially on the east coast between Tanzania and Mozambique. Yara is a leading player in initiatives to promote large-scale industrial agriculture in Africa, such as the World Economic Forum's Southern Agricultural Growth Corridor project in Tanzania, where Yara is coincidentally in talks with the government for the construction of a new $2.5 billion nitrogen fertilizer plant.[45]

Food and climate change: the forgotten link

Graph 2 How fertilizer companies control the Global Alliance
for Climate Smart Agriculture

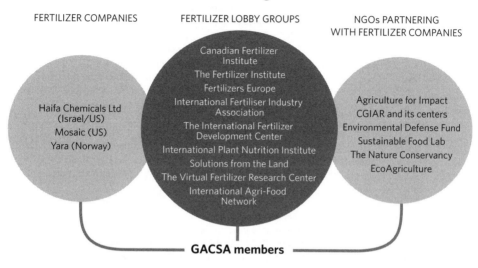

FERTILIZER COMPANIES

FERTILIZER LOBBY GROUPS

NGOs PARTNERING
WITH FERTILIZER COMPANIES

Haifa Chemicals Ltd
(Israel/US)
Mosaic (US)
Yara (Norway)

Canadian Fertilizer
Institute
The Fertilizer Institute
Fertilizers Europe
International Fertiliser Industry
Association
The International Fertilizer
Development Center
International Plant Nutrition Institute
Solutions from the Land
The Virtual Fertilizer Research Center
International Agri-Food
Network

Agriculture for Impact
CGIAR and its centers
Environmental Defense Fund
Sustainable Food Lab
The Nature Conservancy
EcoAgriculture

GACSA members

Original article by GRAIN can be found at https://www.grain.org/e/5270

1 See for example, Gustavo, GT et al (2013) "Energy use and greenhouse gas emissions from crop production using the Farm Energy Analysis Tool". *BioScience* 63(4): 263–273, http://bioscience.oxfordjournals.org/content/63/4/263.full
2 CIDSE (2015) "Climate-smart revolution … or green washing 2.0?", May, http://www.cidse.org/publications/just-food/food-and-climate/climate-smart-revolution-or-a-new-era-of-green-washing-2.html
3 US EIA (2015) "New methanol and fertilizer plants to increase already-growing industrial natural gas use", July, http://www.eia.gov/todayinenergy/detail.cfm?id=22272&src=email. On the shale gas lobby efforts, see: http://shalegas-europe.eu/guest-blog-energy-and-europes-ability-to-create-an-industrial-renaissance-2/?lang=pl and http://www.ifieceurope.org/fileadmin/Downloads/Gas/IFIEC_FE_shale_gas__position_paper_21_02_13.pdf
4 Estimates are from the IPCC. Note that the figures do not include the emissions associated with packaging and transporting fertilizer or the emissions associated with the machinery use to apply them on the farm. See Lin, BB et al (2011) "Effects of industrial agriculture on climate change and the mitigation potential of small-scale agro-ecological farms", CAB Reviews: Perspectives in agriculture, veterinary science, nutrition and natural resources, 6(20), http://www.columbia.edu/~km2683/pdfs/Lin%20et%20al.%202011.pdf
5 FAO (2015) "World fertiliser trends and outlook to 2018", http://www.fao.org/3/a-i4324e.pdf
6 Fischetti, M (2012) "Fracking would emit large quantities of greenhouse gases", *Scientific American*, January http://www.scientificamerican.com/article/fracking-would-emit-methane/ http://www.scientificamerican.com/article/fracking-would-emit-methane/
7 Based on US EPA estimate of 4.7 metric tons of CO_2 per year for the average car driven in the US.

8   See for example, March 2015 study results from Universidad Politécnica de Madrid team showing a 57% reduction in GHG emissions and an 8% increase in yields when urea (nitrogen) fertilizers were removed in "Yield-scaled mitigation of ammonia emission from N fertilization: the Spanish case", *Environmental Research Letters,* http://www.sciencedaily.com/releases/2015/03/150318074403.htm

9   The full report of the IAASTD as well as summaries are available at http://www.globalagriculture.org/report-topics/climate-and-energy.html

10 Two of CFA's six corporate members are Agrium (the world's 9th largest fertilizer company) and the Canadian Fertiliser Institute (the lobby group for the fertilizer industry in Canada and a member of the Global Alliance for Climate Smart Agriculture). See http://www.cfa-fca.ca/about-us/corporate-partners

11 Ifiec Europe and Fertilizers Europe, (2013) Position paper on shale gas, http://www.ifieceurope.org/fileadmin/Downloads/Gas/IFIEC_FE_shale_gas__position_paper_21_02_13.pdf

12 See, for example, the Cool Farm Initiative, which brings together Yara, Unilever, Costco, PepsiCo and the Sustainable Food Lab, another member of the Global Alliance for Climate Smart Agriculture, http://www.coolfarmtool.org/

13 Clayton, C. (2015) "Talking climate-smart agriculture in Abu Dhabi", *Progressive Farmer,* 10 March, http://www.dtnprogressivefarmer.com/dtnag/view/ag/printablePage.do?ID=BLOG_PRINTABLE_PAGE&bypassCache=true&pageLayout=v4&blogHandle=policy&blogEntryId=8a82c0bc49f2d3d3014c04da12fc1365&articleTitle=Talking+Climate-Smart+Agriculture+in+Abu+Dhabi+&editionName=DTNAgFreeSiteOnline

14 FOE International (2013) "Wolf in sheep's clothing: An analysis of the 'sustainable intensification' of agriculture", 2013, http://www.foei.org/wp-content/uploads/2013/12/Wolf-in-Sheep%E2%80%99s-Clothing-summary.pdf

15 Yara (2015) "Pushing climate smart agriculture" Oslo, 16 March, http://yara.com/media/news_archive/pushing_climate_smart_agriculture.aspx

16 See for example, Yara (2014) "Tackling the coffee challenge in Vietnam", September, http://yara.com/media/news_archive/tackling_the_coffee_challenge_in_vietnam.aspx

17 See the Civil Society Statement of Concern on the 2nd Global Conference on Agriculture, Food Security and Climate Change in Hanoi, Viet Nam, 3-7 September 2012, http://www.northchick.org/conservation/agriculture-food-security-climate-change/

18 The Hanoi Communiqué (2012), https://zoek.officielebekendmakingen.nl/blg-188179.pdf

19 Mersmann, C. (2014) "4th partner meeting of the Global alliance for climate smart agriculture", 14 July, https://www.donorplatform.org/cobalt/user-item/660-/3-climate%20change/2262-4th-partner-meeting-of-the-global-alliance-for-climate-smart-agriculture-acsa

20 Personal communication with GRAIN, July 2015

21 CIDSE (2015) "Climate-smart revolution ... or green washing 2.0?", May, http://www.cidse.org/publications/just-food/food-and-climate/download

22 GRAIN (2009) "Earth matters – Tackling the climate crisis from the ground up", 28 October, https://www.grain.org/e/735

23 GRAIN (2008) "Making a killing from hunger", April, https://www.grain.org/e/178

24 Robert Sanders (2012) "Fertiliser use responsible for increase in nitrous oxide in atmosphere", *Berkeley News,* 2 April, http://news.berkeley.edu/2012/04/02/fertilizer-use-responsible-for-increase-in-nitrous-oxide-in-atmosphere/

25 Creutzen et al (2008) "N$_2$O release from agro-biofuel production negates global warming reduction by replacing fossil fuels", *Atmos. Chem. Phys.,* 8: 389–395, http://www.atmos-chem-phys.net/8/389/2008/acp-8-389-2008.pdf

26 Mulvaney et al (2009) "Synthetic nitrogen fertilizers deplete soil nitrogen: A global dilemma

for sustainable cereal production", *J. Environ. Qual.* 38: 2295–2314, https://dl.sciencesocieties. org/publications/jeq/pdfs/38/6/2295

27 Shcherbak et al (2014) "Global metaanalysis of the nonlinear response of soil nitrous oxide (N_2O) emissions to fertilizer nitrogen", *PNAS*, January, http://www.pnas.org/ content/111/25/9199.full

28 Schepers, JS and Raun,WR (ed) (2009) "Nitrogen in agricultural systems", *Agron. Monogr 49*

29 GRAIN (2009) *op.cit.*

30 Khan et al (2007) "The myth of nitrogen fertilisation for soil carbon sequestration", *J. Environ. Qual* 36:1821–1832 , http://www.ncbi.nlm.nih.gov/pubmed/17965385

31 Science Daily (2007) "Nitrogen Fertilizers Deplete Soil Organic Carbon", 30 October, http:// www.sciencedaily.com/releases/2007/10/071029172809.htm

32 GRAIN (2009) *op.cit.*

33 CIDSE, *op.cit.*

34 Global Alliance for Climate Smart Agriculture, brochure, http://www.fao.org/3/a-au980e.pdf

35 FAO (2013) "Success stories on climate smart agriculture", http://www.fao.org/3/a-i3817e.pdf

36 CGIAR (2013) "Climate smart agriculture success stories with farming communities around the world", https://cgspace.cgiar.org/bitstream/handle/10568/34042/Climate_smart_ farming_successesWEB.pdf

37 ISID (2014) "CIMMYT/IPNI fertiliser efficiency tool wins global innovation prize", Climate change policy and practice, 20 February, http://climate-l.iisd.org/news/ cimmytipni-fertilizer-efficiency-tool-wins-global-innovation-prize/

38 US Department of Agriculture (2015), "Agriculture and forestry: Part of the climate solution", http://www.usda.gov/wps/portal/usda/usdahome?contentidonly=true&contentid=climate-smart.html

39 EDF (2015) "Campbell Soup Company joins EDF initiative to reduce environmental impact of food production", *Oklahoma Farm Report*, 2 June, http://oklahomafarmreport.com/wire/ news/2015/06/09482_CampbellJoinsEnvironmentalDefenseFund060215_114522.php

40 Walmart, "Sustainable food", http://corporate.walmart.com/global-responsibility/ environment-sustainability/sustainable-agriculture;

41 Gunther, M. (2015) "Walmart targets climate-smart suppliers", *Corporate Knights*, 24 April, http://www.corporateknights.com/channels/food-beverage/ walmart-targets-climate-smart-suppliers-2-14298636/

42 Yara, "Global fertilizer brands", http://yara.com/products_services/fertilizers/global_brands/ yaraliva.aspx

43 Yara, "Reducing carbon footprints", http://yara.com/media/stori es/tropicana_carbon_ footprint_project.aspx

44 Yara (2015) "Pushing climate smart agriculture", Oslo, 16 March, http://www.yara.com/ media/news_archive/pushing_climate_smart_agriculture.aspx

45 Nachilongo, H. (2014) "Norwegian firm sets aside $2.5b to build gas, oil plants", *The East African*, 27 September, http://www.theeastafrican.co.ke/business/Norwegian-firm-sets-aside--2-5b-to-build-gas--oil-plants-/-/2560/2467020/-/pv6qml/-/index.html

1.4 How REDD+ projects undermine peasant farming and real solutions to climate change

Peasants do an incredible job of providing most of the world's food on just one quarter of the world's agricultural lands. But ask any of these small-scale farmers about climate change and they will tell you how it is making farming more difficult. It is getting harder for them to predict the weather, while storms, floods and droughts are becoming more frequent and extreme.

Scientists and politicians have begun to acknowledge the threat that climate change poses to global food security and some have come around to the reality that industrial agriculture is a major contributor to climate change. Agriculture is increasingly being discussed at high-level forums on climate change, and governments and international agencies are coming forward with different programs that they claim will help farmers to adapt to climate change and mitigate agriculture's greenhouse gas emissions.

These various initiatives are all politically loaded, just like any other area of international agricultural policy. They are heavily influenced by powerful corporations and governments that want to protect industrial agriculture and corporate food systems from real solutions to climate change, which would provide peasants with more lands and support agro-ecological farming for local markets. As a result, small-scale peasant agriculture is being targeted by a number of aggressively promoted false solutions to climate change while industrial and corporate-driven agriculture mostly continues, business as usual.

In this context peasant organizations are under increasing pressure from NGOs, governments and donors to engage their members in new programs on small-scale farming and climate change. There are growing numbers of workshops, booklets and handbooks that promote initiatives with awkward names, such as REDD+ (*See Box 1: What is REDD+? on page 28*) or climate smart agriculture. In addition, many industrialized countries and international conservation groups are funding pilot REDD+ projects aimed at peasant farmers. Although these initiatives all claim to benefit small farmers, the reality is that most undermine peasant farming and food systems by claiming that traditional agricultural practices, especially shifting cultivation, are major cause of climate change and forest loss and by robbing peasants of access to lands and forests or restricting what peasants can do on their lands.

REDD (**R**educing **E**missions from **D**eforestation and Forest **D**egradation) is advertised as a solution that can help peasants reduce emissions, adapt their farming practices to a changing climate and increase yields. Despite promising wins for all sides, experience has shown that REDD+ is not an ally of peasant communities. In 2014, the World Rainforest Movement (WRM) compiled reports about 24 existing REDD+ initiatives, "REDD: A Collection of Conflicts, Contradictions and Lies".[1] It revealed that in most cases, the information peasant communities had received about REDD+ projects was biased or incomplete. Many promises of benefits and employment were made by project proponents on the condition that communities agreed to the proposed REDD+ activity. What the villagers got in return for the promises, however, was mainly harassment, loss of access to land and blame for being responsible for deforestation and causing climate change.

Almost all REDD+ activities limit the use of the forest for shifting cultivation, gathering and other subsistence use. Hunting, fishing, grazing or cutting some trees for construction of housing or canoes are also often restricted and the restrictions are enforced by REDD+ project owners, often with the support of armed guards. At the same time, large-scale drivers of deforestation, such as industrial logging, expansion of oil palm, soya or tree plantations, infrastructure mega-projects, mining, large hydro dams – and above all, industrial agriculture expanding into the forest – continue without restriction.

In very few of these cases are communities informed that the 'product' these REDD projects generated, the carbon credits, will be sold to polluters in industrialized countries. That the buyers of these carbon credits are some of the largest corporations worldwide, whose businesses are built on fossil fuel extraction and the destruction of the territories of indigenous peoples and forest communities is rarely revealed. Indeed, in the vast majority of these REDD+ projects, peasant farming is singled out as the cause of deforestation while the major drivers of deforestation – extraction of oil, coal, mining, infrastructure, large-scale dams, industrial logging and international trade in agricultural commodities – are ignored.[2]

REDD+ is not just a false solution to the urgent and critical problem of climate change. It reinforces the corporate food and farming system that is largely responsible for climate change, has robbed many communities and forest peoples of their territories and undermines the food and agricultural systems of peasants and indigenous peoples that can cool the planet.

Box 1: What is REDD+?

REDD stands for Reducing Emissions from Deforestation and Forest Degradation in developing countries. It is the term under which forest loss is discussed at United Nations (UN) climate meetings. Since 2005, the issue of forest loss has distracted governments at these UN meetings from addressing the real cause of climate change – turning ancient underground deposits of oil, coal and gas into fossil fuels and burning them. Instead of coming up with a plan on how to end the release of greenhouse gas emissions that is the consequence of burning these fossil fuels, the UN climate talks have spent much time debating deforestation of tropical forests. Of course it is important to halt forest loss, also because of the CO_2 emissions that are released when forests are destroyed. But reducing deforestation is no substitute for coming up with a plan on how to stop burning fossil fuel! The trouble with REDD is that that is exactly its consequence: enabling industrialized countries to burn fossil fuels a little longer.

REDD+ is another word the UN uses to discuss forests, and the plus stands for "enhancing carbon stocks, sustainable forest management and forest conservation" – or, as one commentator stated, "at some stage someone thought it fitting to tag on the '+' which would come to represent all those other things that have come to the attention of the international development industry in recent years (like conservation, gender, indigenous people, livelihoods and so on)". REDD was originally designed for countries with high deforestation, Brazil and Indonesia in particular. This meant that funding would be available primarily for those countries with much potential to reduce their rate of deforestation. Only eight countries, accounting for 70% of tropical forest loss, would thus be involved. Countries with much forest but little deforestation – Guyana, DRC, Gabon, for example – therefore insisted that REDD be designed so they would also have access to REDD funding, for example through being paid to not increase projected future deforestation. The plus was thus added so that countries with low levels of deforestation but a lot of forest could also have access to what was at the time expected to be large sums of money for REDD+ activities.[3]

How is REDD+ meant to work?

Forest-rich countries in the Global South agree to reduce emissions from forest destruction as part of a UN climate agreement. To demonstrate exactly how many tonnes of carbon (dioxide) have been saved, the government produces a national REDD+ plan, which explains how much forest *would have been* destroyed over the next few decades. Then they describe how much forest they would be willing not to cut if someone paid them to keep the forest standing. They calculate how much it would cost not to destroy this forest and how much carbon will not be released into the atmosphere as a result of keeping the forest intact.

In return, industrialized countries (or companies or international NGOs) pay the tropical forest countries (or individual REDD+ projects) to prevent the forest destruction that is claimed to happen without REDD+ finance. The payment will only be made if the forest country shows that forest loss has actually been reduced *and* that the carbon that otherwise would have been released into the atmosphere continues to be stored in the forest. That is why people sometimes talk about "results-based" or "performance" payments for REDD+. The REDD+ project also needs to show that without the REDD+ money the forest would have been destroyed. This last point is important because many industrialized countries and corporations that fund REDD+ activities want to receive something in return for their financial support. This something is called a *carbon credit*. The WRM publication, "10 things communities should know about REDD"[4], explains why the calculations that create carbon credits are not credible and why it is impossible to know whether forest was really only saved because of the REDD+ money.

What is this carbon credit good for?

A carbon credit is essentially a right to pollute. A polluting country or company that has made a commitment to reduce its greenhouse gas emissions does not reduce its emissions by as much as it said it would. Instead, it pays someone elsewhere to make the reduction for them. That way, the polluter can claim to have lived up to their commitment when in reality they continue burning more oil and coal and release more CO_2 into the atmosphere than they said they would. At the other end of the (REDD+) carbon credit deal, someone claims they were planning to destroy a forest but as a result of the payment, they decided to not destroy that forest. The carbon saved by protecting the forest that otherwise would have been cut is sold as a carbon credit to the polluter who keeps burning more fossil fuels than agreed. In other words, the owner of the carbon credit has the right to release one tonne of fossil carbon they had promised to avoid because someone else has saved a tonne of carbon in a forest that without the carbon payment would have been destroyed, releasing CO_2. On the voluntary carbon market, where corporations and individuals buy carbon credits to claim that (some of) their emissions have been offset, REDD+ credits are traded for between $3 and $10.

Why does trading carbon credits not reduce emissions?

There are many problems with this idea of (carbon) offsets. Among them that they do not reduce overall emissions – what is saved in one place allows extra emissions in another place. In the case of REDD+ offsets, another problem is the very important difference between the carbon stored in oil, coal and gas and the carbon stored in forests. The carbon stored in the trees is part of a natural cycle through which carbon is constantly released and absorbed by plants. The terrestrial carbon has been circulating between the atmosphere, the oceans and the forest for millions of years.

Deforestation over the centuries has meant that too much of the carbon naturally in circulation has ended up in the atmosphere and too little in forests. Today, industrial agriculture, logging, infrastructure and mining are the main drivers of deforestation. When industrialized countries started burning oil and coal, they further increased the amount of carbon that could accumulate in the atmosphere. The carbon in these "fossil fuels" had been stored underground for millions of years, without contact with the atmosphere. Its release greatly increases the amount of carbon dioxide in the atmosphere, which in turn causes the climate to change. Although plants can absorb part of this additional carbon released from ancient oil and coal deposits, they do so only temporarily. When the plant dies or a forest is destroyed or burns, the carbon is released and increases the concentration of CO_2 in the atmosphere (adding to the imbalance from forest destruction).

That is why REDD+ credits not only don't help reduce overall emissions, but also will lead to an increase of CO_2 concentrations in the atmosphere, because REDD+ is built on the false assumption that forest and fossil carbon are the same when from a climate perspective they are clearly not!

Patterns that make REDD+ a danger to peasant farming

REDD+ blames peasant farming practices for deforestation and emissions

Peasants around the world are being squeezed onto less and less land. Today, they account for 90 per cent of the farms, but occupy only one quarter of the world's agricultural lands. Yet they still manage to produce most of the world's food, without nearly the amount of GHG emissions produced by large-scale industrial farms. Any program that would take more land away from peasant communities can therefore not be a solution to the climate crisis. To cool the planet, the world needs more small farmers farming on a greater percentage of the world's agricultural lands, and less land in the hands of big corporate farms.

The overwhelming majority of REDD+ projects, however, seek to reduce GHG emissions by further reducing the lands that peasant farmers and indigenous communities have access to or by changing how the land is used.[5] REDD+ proponents justify their backwards approach with the erroneous assumption that shifting cultivation in particular, a practice commonly used by peasants around the world, is a major cause of deforestation. This is simply not true.

Shifting cultivation is a land-use practice that peasants have developed over

31

many generations of growing food in challenging conditions. What is usually lumped together under the term "slash-and-burn" in reality are hundreds of different land-use practices, adapted to the local circumstances. Far from causing large-scale forest loss, these practices have allowed forest-dependent communities to maintain the forests they depend on.

A recent CIFOR report on the Democratic Republic of Congo, for example, found a "lack of strong evidence" that peasant agriculture contributed significantly to overall deforestation and concluded that any biodiversity and carbon impacts from deforestation by peasants would be limited.[6] Another recent study of coastal Madagascar pointed to historical droughts as a cause of deforestation rather than peasant farming or shifting cultivation, as has been widely assumed.[7] Where shifting cultivation is leading to forest degradation, rotation cycles are usually shortened because less land is available for shifting cultivators. This is almost always a result of expanding industrial plantations or mega-infrastructure projects or industrial logging, which grab land peasant communities rely on for food production.

Another erroneous assumption used by REDD+ proponents that justifies their focus on peasant practices is that the "opportunity cost" is lower than it is with restricting the expansion of plantations and industrial farms. The "opportunity cost" equals the cost of not cutting down forests. It is a measure of the economic value that would have been generated, by companies or peasants, if deforestation activities were allowed to continue. But, in the biased eyes of the consultants hired on REDD+ projects, the economic costs of not proceeding with a plantation are much higher than the costs of not proceeding with the planting of a local food crop by peasants or the costs of restricting a community's access to the forest for hunting and gathering or for grazing. The consultants can see the money that plantations generate for companies; but they do not see the whole value that forest areas represent for peasant communities in terms of local food production, housing, medicines, biodiversity and culture, for example. For REDD+ proponents, therefore, it is more "cost"-effective to stop peasants from using forestlands than it is to stop plantation companies and corporate farmers.

This approach suits the industrialized countries and international aid agencies that fund most REDD+ projects. It means that for relatively little money they can present the image to the world that they are doing something about deforestation without having to address their own responsibility for deforestation, through the promotion and consumption of industrial agriculture products for export.

REDD+: Good business for carbon companies, international conservation NGOs, consultants and industrialized countries

One of the big promises of REDD+ is that forest-dependent communities and peasant farmers will get paid for protecting the forest. To entice governments and communities of the South, REDD+ proponents routinely make exaggerated claims about the size of the global trade in carbon credits – or the expected size of a future forest carbon market.

"Imagine a market that could provide billions of dollars for replanting trees, protecting standing forests, and improving the way timber is harvested. That is what we are talking about when we talk about the potential of carbon markets, and the role forest carbon might play in them."[8] This is how Mark Tercek, the CEO of US-based conservation group The Nature Conservancy, one of the strongest proponents of REDD+, described the potential of carbon markets for forests at a Carbon Finance Speakers event at Yale University in 2009.

In 1997, when the UN's international climate treaty, the Kyoto Protocol, allowed industrialized countries to achieve their emission limits in part by paying for reductions in the Global South, similar promises were made. The World Bank and the same international conservation groups that today advocate for a forest carbon market predicted that the Kyoto Protocol's Clean Development Mechanism (CDM) could bring billions to the poor in the Global South. But, today, just a few ailing regional carbon markets are all that has materialized from the projected multi-billion, if not trillion dollar global carbon market that was supposed to pave the way for carbon to become the world's new global currency.

The reality is that the price for carbon permits has been in free fall since 2008, among other reasons because governments gave out so many permits to companies for free that few companies needed to buy more permits to cover their emissions. Emission permits in the largest carbon market, the EU Emissions Trading Scheme, now trade at around €7 – far below the €42 level that would be needed to encourage German utilities to switch from burning coal to natural gas and even further from the €60-€80 price that these permits were predicted to trade at when the scheme was introduced. Carbon credits from CDM projects are faring even worse and have been trading for as little as €0.40 for a few years now. In fact, the financial performance of carbon markets is so bad that the World Bank stopped issuing its annual State of the Carbon Market Report in 2012, when it could no longer find a way to at least show some positive development in carbon markets.

Carbon permits might swing back to the expected price. But the experiences of existing REDD+ projects that sell carbon credits in the voluntary carbon market, where corporations and individuals buy carbon credits to claim that (some of) their emissions have been offset, show how most of the supposed profits that are in theory going to communities will be captured by others.

Before a REDD+ project can sell carbon credits, a lot of technical documents have to be written, certified and verified by different auditing firms.[9] Most of the time, the REDD+ project also needs the help of middlemen to find buyers for its credits. This is always the case in those rare situations where a community itself runs the REDD+ project. All of these preparations do not only use a jargon language, but they also cost money. And they are not cheap. They add up to what is called the "overhead costs" or "transaction costs" of REDD+ projects. These vary from case to case but typically they are between 20 per cent and 50 per cent of the offset project budget. Payments to communities are also usually of net, not gross, profit – and anecdotal experience suggests that there often is not much net profit left after the project owners have deducted all their costs.

For international conservation groups like The Nature Conservancy, Conservation International and WWF by contrast, REDD+ is good business because they are able to capture a large portion of the international aid and climate funding available for REDD+. They are involved in many REDD+ projects and initiatives and act as advisors on national REDD+ plans. None of these groups have revealed the size of their REDD+ budgets, or how much of their funding comes from the climate finance that industrialized countries account as REDD+ payments to the Global South.

Communities participating in REDD+ projects can also be saddled with financial risks and obligations contained in their contracts, which were often not clearly explained to them. For example, in one tree planting project in Ecuador, run by the Dutch company FACE, the carbon contract between the company and the participating communities included an obligation for the community to replant trees that might be destroyed, for example in wild fires. The trees planted were pine trees in monoculture plantations and in a region that is not suitable for pine and has a high risk of fires. It was therefore not really a surprise when the carbon trees burned down – in one location not once but three times! The first time, the community paid to have the trees replanted because the company insisted on fulfilment of the contract obligations. But when the trees burned down again, they refused to pay and the company threatened to take legal action against them.[10]

Industrialized countries also stand to gain even more from REDD+ if the new UN climate treaty currently being negotiated provides them with the possibility to take the credit for tropical countries reducing deforestation. A decision on how reducing forest loss will be financed under a new UN climate treaty is expected from the UN climate meeting in Paris in December 2015. One of the proposals on the table is that the countries providing financial support for REDD+ count REDD+ reductions towards their own emission targets. If the country where deforestation was reduced does the same, the same reduction would in effect be claimed twice, resulting in actual emissions of greenhouse gases being higher than reported to the UN. Therefore, if tropical forested countries cannot agree to industrialized countries taking the credits for their REDD+ emissions reductions, they should not agree to REDD+ being funded by an international trading mechanism.[11]

REDD+ undermines food sovereignty

There are different ways that REDD+ projects commonly undermine local food production and create food insecurity among local communities. In some cases, families participating directly in the offset project must reduce their production of food crops in order to plant trees for the project. In other cases, the REDD+ project prevents the communities from accessing forested areas that they rely on for hunting and gathering, for shifting cultivation or for grazing.

Because most REDD+ projects start from the false assumption that shifting cultivation and peasant farming in forest areas are a threat to both forests and the climate, they generally include restrictions on families opening new fields in the forest. The documents usually include proposals to increase yields on existing plots, through "modernizing" practices, such as intercropping to maintain nutrients and soil fertility. The reality, however, is that the large majority of these proposals fail because they are not suitable for the particular local circumstances.

The experience that a community had in Bolivia with a forest carbon offset project is typical of REDD+ projects elsewhere. A villager from the community told researchers about a herd of cows the offset project had provided in an attempt to set up "alternative livelihoods" for the community to make up for the loss of access to forested lands. Unfortunately, the cows were European breeds, unable to survive in Bolivia. "They all died in the end," the villager said. "The cows were so expensive that a whole herd of local breeds could have been bought for the price of a single one."[12]

The regular failure of such attempts to establish alternatives to slash-and-burn or "modernize" peasant agriculture through proposals developed by far-away REDD+ project owners or conservation NGOs points to another tension inherent in REDD+: these projects are concerned first and foremost with maximizing carbon storage in the area that will deliver carbon credits. Initiatives to involve peasant communities and forest peoples are an afterthought, a requirement from donors or to show participatory project implementation.

Hardly ever are the needs of forest-dependent communities the genuine starting point for designing such projects. Consequently, failure of initiatives aimed at increasing crop yield or developing new income generation opportunities is predictable for local participants. The ideas might sound good on paper but regularly fail to reflect local circumstances.

REDD+ undermines community control over territories

"Why remain in the forest, if you are forbidden to live with it?"

Dercy Teles, Rural Workers Trade Union, Xapuri, Acre

Tradable REDD+ credits are a form of property title. Those who purchase the credits do not need to own the land or the trees that are "storing" the carbon, yet they own the right to decide how that land will be used. They also usually have contractual rights to monitor what is happening on the land and request access to the land at any time they choose for as long as they own the carbon credit.

Communities often are not informed about how the contract they sign for REDD+ projects might undermine their control over their territories. In 2013, Friends of the Earth International analyzed a number of REDD+ project contracts that involved communities directly and found that many

REDD+ contracts were full of "words written with the intention of not being understood, not being fulfilled". Often, obligations that communities or families enter into are not clearly explained or they are described in ambiguous terms that can easily be misinterpreted. Seeking legal advice on such complex and ambiguous technical documents is made difficult because almost all REDD+ contracts contain strict confidentiality clauses. Many of the contracts and project documents are also written in English, with only partial or no translation into local languages, which further restricts the possibility for communities to fully inform themselves about the REDD+ projects presented to them.

Community control over territories is also undermined by the inbuilt logic of carbon offsets, which requires that the REDD+ project identifies the users of the land and their activities as a threat to the forest in order for the REDD+ project to generate carbon credits. If the activities are not a threat to the forest, there is no risk of deforestation and therefore no carbon credits can be generated from avoiding deforestation! For REDD+ projects involving forest communities this means that people who for generations have protected the forest must describe the way they use the forest as a risk in the hypothetical story of what would have happened with the forest without the REDD+ project. Without such a story that the forest would have been destroyed, there is no carbon to be saved, and thus no carbon credits to be sold. This requirement of the REDD+ offset project to describe peasant farming and shifting cultivation as a risk to the forest is already reinforcing the dangerous false belief that forest-dependent communities and small-scale farmers are the most important agents of deforestation and undermines the control these communities have over their territories.

Another important way that REDD+ projects affect community control over territories is by creating divisions within communities. While many promises of employment through REDD+ projects remain unfulfilled, REDD+ projects generally hire people from within the community to work as forest rangers or guards, whose role it is to report on compliance with REDD+ project rules within the community. In other words, they are expected to keep an eye on other members of the community. Their role is to report to the project owners if community members cut down trees, hunt, fish, grow food crops in the forest or use the forests as they have always done but which is forbidden under the REDD+ project rules.

Needless to say, this is a job prone to creating conflict within the community, in particular if the rules were not agreed with the community but imposed by the REDD+ project. This form of employment creates divisions within the community that will negatively affect the ability of communities to organize and work together to defend their territories.

Box 2: "What have we gained? Not much"

In 2002, the N'hambita Community Carbon Project in Mozambique was started with a €5 million EU grant by Envirotrade, a company registered originally in Mauritius. The aims of the project included conserving a community-owned forest, introducing agroforestry and other new farming practices to improve crop yields, and establishing community enterprises. Local people were contracted to plant and care for trees on their land, and communities were also tasked with protecting and patrolling a 10,000ha forest area. Opening new fields was not allowed. The project did initially provide some income for people and allowed some families to put tin roofs on their houses or buy a solar panel and run a little business to charge phones etc. But these benefits pale in comparison with the long-term legal obligations involved. Villagers are paid for seven years to plant and conserve trees, but sign a contract for 99 years. "It is the farmer's obligation to continue to care for the plants which they own, even after the seven-year period covered by this contract", a clause in the contract states. António Serra from Envirotrade in Mozambique told La Via Campesina who investigated the project in 2012: "If a farmer passes away during the contract period, the contract, all the rights contained therein but also all the obligations, are transferred to their legitimate/legal heirs." When the researchers examined one farmer's contract they found that he would be paid $128 over seven years for planting trees in an area of 0.22ha.

At these kinds of rates the farmer would need to have access to a much greater area of land than most farmers in the community had and would have to plant many more trees to "alleviate poverty" – another stated project objective. The payments to farmers are also conditional upon 85 per cent of the seedlings surviving – otherwise payments are reduced.

As a consequence, many villagers involved in the project reduced or stopped farming so they could tend the trees. But still, regularly less than the required 85 per cent of the seedlings survived. When payments were reduced or delayed, the lack of money combined with having given up or scaled back farming made their already difficult situation worse.

A report for La Via Campesina also found that a considerable number of farmers involved in maintaining firebreaks and patrolling the community forests in the REDD+ area had abandoned farming. One villager who co-ordinated a group of farmers maintaining firebreaks and patrols used to farm to feed his family. "Now our main activity is firebreaks. I don't have time to go to the machamba," he says.[13] The $340 he earned during the firebreak season he has to divide between the group of four that he manages. Securing food has thus become more difficult for many involved in the project.

Box 3: "I and my people have suffered for five years now"

In Cross River State, southeast Nigeria, a REDD+ program that involves the FAO, UNDP and UNEP includes a moratorium on forest activities that community members have depended on for generations.

"I and my people have suffered for five years now since government stopped us from entering our forest because REDD is coming and till now I have not received anything from them," says Chief Owai Obio Arong of Iko Esa Community.

Under the program, products like kola nuts or fruits deemed to have been collected from the REDD+ forest area are confiscated from community members. The harvesting of Afang leaves, a local vegetable consumed in West and Central Africa, has also been banned in forests designated by the government as REDD+ areas. This criminalization of food gathering from the forests and related economic activities has promoted an underground market which in turn has driven up the price of forest products. The REDD+ program has essentially turned community forests into state-controlled areas.[14]

Box 4: "Suffering here to help them over there"

The Nature Conservancy's Guaraqueçaba Climate Action Project in the south of Brazil is one of the early forest carbon projects. In promotional materials, the project owners write that it is important "to ensure that local people had a stake in keeping the forests around Guaraqueçaba standing. Everyone has to make a living somehow - so if you can't farm or ranch, how can your family earn money? That's why we and our partners have involved so many community members in income-generating, sustainable enterprises."

The income-generating, sustainable enterprises and the employment the project provided were short-lived. What remained, however, were restrictions on traditional communities' use of their territories, including the forests they had protected for generations. Harassment of people entering the forests to gather food, wood, or vines became ever more frequent, and many families started to move away from the place that was their home.

"Directly or indirectly, it was through these conservation projects that the population came here and created a ring of poverty around our city, causing a really big social problem here," the mayor of the nearby town Antonina explains in a film about the project. "It's a game that only has economic aims. It favors big businesses and NGOs. They don't care about the environment, they care about profit, the NGOs as much as the businesses; through carbon credits, they keep polluting, they keep earning more. And it's the community that pays the price for all of this."

REDD+ facilitates the expansion of corporate agriculture

The deforestation caused by the agriculture sector over the past few decades is almost entirely due to the expansion of commodity crops for export and for animal feed. The land occupied for growing just four of these crops – soybean, oil palm, rapeseed and sugar cane – has quadrupled over the past five decades, and the vast majority of this expanded production is on large-scale industrial farms and plantations.[15,16,17]

Deforestation is therefore directly linked to the international commodity supply chains that are controlled by a small number of transnational food corporations. These include commodity traders and producers such as Cargill, Louis Dreyfus Group, Bunge, Archer Daniels Midland (ADM), JBS or Wilmar International, food companies such as Nestlé, Danone, or Unilever, and supermarkets and fast-food chains such as McDonald's, Walmart or Carrefour.[18]

To shield themselves from bad publicity and to protect their supply channels, corporations have established voluntary certification schemes and commodity roundtables with the participation of a few large international NGOs. Such roundtables now exist for timber products (FSC), palm oil (RSPO), soya (RTRS), sugar (Bonsucro) and beef (BRBS). All these initiatives have developed a set of standards against which producers are certified, usually by third-party auditors paid by the enterprise seeking certification, and which have been criticized for greenwashing corporate destruction and failing to address the issue of overconsumption.[19,20,21,22]

In the past few years, the connections between these commodity roundtables, certification schemes and initiatives linked to deforestation, climate change and REDD+ have been increasing. All the major roundtables now include requirements related to greenhouse gas emissions, such as identifying "high carbon value forests", exploring carbon accounting methods, working towards "zero deforestation" commodities or engaging in carbon offsetting initiatives. With this increasing merger of commodity roundtables and "zero deforestation" initiatives, the focus of REDD+ has expanded from forests to so-called "landscapes". From late 2013 onwards, terms like "landscape REDD", "landscape funds", or "land-scape investment" have been increasingly mentioned in the same breath with REDD+.

The World Bank plays a key role in bringing "landscape" initiatives and REDD+ together with carbon markets. On the sidelines of the 2013 UN climate meeting, Norway, the United Kingdom, and the US together committed $280 million for the bank to set up the Initiative for Sustainable Forest Landscapes (ISFL) as part of its already existing BioCarbon Fund. The BioCarbon Fund is a public-private partnership, housed in the World Bank; it was the first carbon fund to implement carbon offset projects in the forest and agriculture sector. Unilever, Mondelēz International and Bunge were among the food corporations involved in the preparation of the ISFL and were present at the launch of the initiative. The World Bank announced its new Initiative for Sustainable Forest Landscapes with the promise of "creating multiple revenue streams from the sustainable transformation of landscapes".

This merger of REDD+ and agricultural commodity production provides huge opportunities for transnational food corporations, such as Unilever and Cargill, to protect their "revenue streams" and even create new ones. Both companies are members of the Consumer Goods Forum, a "collaboration of 400 retailers, manufacturers, and service providers with combined annual sales of over $3 trillion" that have committed to move towards a goal of zero net deforestation in their supply chains by 2020. What "zero net defor-estation" really means is that companies can continue to source agricultural commodities from deforested areas as long as trees are planted in compen-sation or forests elsewhere are protected by REDD+ programs. It means that corporations get control over forests (to use for commodity production) and peasant communities and indigenous peoples lose control over forests and can no longer use them for food production or their livelihoods. Under the "landscape REDD" scenario, whole territories would be parcelled out by companies into forested areas that provide them with carbon credits and farming areas where they would set up plantations and force local farmers into contract production arrangements.

The problems are clear, the solutions exist

The big gap between the reality and the promises of the REDD+ promoters shows that, for peasants, REDD+ is a false solution that undermines food sovereignty and the control forest-dependent communities have over their lands. REDD+ also helps to conceal the fact that while agriculture is a major contributor to climate change, not everybody growing crops shares the same responsibility for the emissions. It is the industrial food system – with its heavy use of chemical inputs, its erosion of soils, its deforestation and its emphasis on production for export markets – that is the main source of greenhouse gas emissions.

In reality, peasants are already proving that it is possible to "feed the world" while producing far fewer emissions than the export-led, industrial model of agricultural production. Giving lands back to small farmers and indigenous communities is the most effective way to deal with the challenges of feeding a growing global population in an era of unpredictable climate change. REDD+ is a dangerous distraction from urgent action in this direction.

This chapter was extracted from a booklet jointly published by GRAIN and the World Rainforest Movement. The full booklet can be dowloaded from: www.grain.org/e/5322

Food and climate change: the forgotten link

1 http://wrm.org.uy/books-and-briefings/
redd-a-collection-of-conflicts-contradictions-and-lies/

2 See for instance the now cancelled Kalimantan Forest Climate Partnership, described in Yayasan Petak Danum Letter to the Australian Delegation to Central Kalimantan February 2011, RE: Community Concerns with the KFCP, http://www.redd-monitor.org/wp-content/uploads/2011/02/YPD-Letter-to-Australian-Delegation.pdf

3 For more information, see WRM website section on REDD and publication "10 Things Communities Should Know About REDD", www.wrm.org.uy

4 See: http://wrm.org.uy/articles-from-the-wrm-bulletin/recommended/10-things-communities-should-know-about-redd-re-launched-with-a-new-introduction/

5 WRM (2015) " REDD: A Collection of Conflicts, Contradictions and Lies", http://wrm.org.uy/wp-content/uploads/2014/12/REDD-A-Collection-of-Conflict_Contradictions_Lies_expanded.pdf

6 Ickowitz A et al. (2015) "Agriculture and deforestation in the Democratic Republic of the Congo: A synthesis of the current state of knowledge", Occasional Paper 119. Bogor, Indonesia: CIFOR, http://www.cifor.org/publications/pdf_files/OccPapers/OP-119.pdf

7 Virah-Sawmy, M. (2009) "Ecosystem management in Madagascar during global change", Conservation Letters, 2: 163–170

8 Tercek, M. (2009) "Protecting Forests and Lands through Environmental Markets and Finance", Carbon Finance Speakers Series at Yale, 10 February, p.35

9 See the Climate, Community & Biodiversity Alliance website for examples of what such documents look like. They are rarely less than 100 pages long! http://www.climate-standards.org/category/projects/

10 Yanez, I. (2015) "Josefina and the Water Springs against Pine Plantations in Ecuador's Páramos", WRM Bulletin 211, March, http://wrm.org.uy/articles-from-the-wrm-bulletin/section1/josefina-and-the-water-springs-against-pine-plantations-in-ecuadors-paramos/

11 FERN & TWN (2015) "Who takes the credit? REDD+ in a post-2020 UN climate agreement, http://www.fern.org/sites/fern.org/files/Who%20takes%20the%20credit.pdf

12 Greenpeace (2009) "Carbon Scam: Noel Kempff Climate Action Project and the Push for Subnational Forest Offsets", http://www.greenpeace.org/usa/Global/usa/report/2010/1/carbon-scamnoel-kempff-clima.pdf

13 La Via Campesina Africa (2012) "Carbon trading and REDD+ in Mozambique: farmers 'grow' carbon for the benefit of polluters" https://www.grain.org/bulletin_board/entries/4531-carbon-trading-and-redd-in-mozambique-farmers-grow-carbon-for-the-benefit-of-polluters#sdfootnote2anc

14 Social Development Integrated Centre (2014): "Seeing REDD. Communities, Forests and Carbon Trading in Nigeria". http://www.rosalux.sn/wp-content/uploads/2011/02/SEEING-REDD-ready- 1-version-new.pdf

15 GRAIN (2014) "Hungry for land: small farmers feed the world with less than one–quarter of all farmland", https://www.grain.org/article/entries/4929-hungry-for-land-small-farmers-feed-the-world-with-less-than-a-quarter-of-all-farmland

16 Persson, M. et al. (2014) "Trading Forests: Quantifying the Contribution of Global Commodity Markets to Emissions from Tropical Deforestation", CGD Working Paper 384, http://www.cgdev.org/sites/efault/files/CGD-Climate-Forest-Series-8-persson-et-al-trading-forests_0.pdf

17 Hosonuma, N., et al. (2012) "An assessment of deforestation and forest degradation drivers in developing countries", Environmental Research Letters, Vol 7; Also see Forest Trends (2014) "Consumer Goods and Deforestation", September, http://www.forest-trends.org/documents/files/doc_4719.pdf

18   Hosonuma, N. et al. (2012), *Ibid.*

19   WRM (2010) "RSPO: The 'greening' of the dark palm oil business, http//www.rspo-the-greening-of-the-dark-palm-oil-business/

20   Overbeek W. et al (2012) "An overview of industrial tree plantation conflicts in the global South. Conflicts, trends, and resistance struggles", EJOLT Report No. 3, http://www.ejolt.org/2012/06/an-overview-of-industrial-tree-plantations-in-the-global-south-conflicts-trends-and-resistance-struggles/

21   WRM (2013) "FSC consultation and complaints procedures: the case of Veracel Celulose in Brazil", http://wrm.org.uy/books-and-briefings/new-briefing-on-fsc-certification-of-plantations/

22   WRM (2013) "12 Replies to 12 Lies about Oil Palm monocultures plantations", http://wrm.org.uy/books-and-briefings/12-replies-to-12-lies-about-oil-palm-monocultures-plantations/

1.5 Trade deals boosting climate change: the food factor

The climate talks in Paris in December 2015 were viewed as a last chance for the world's governments to commit to binding targets that might halt our march towards catastrophe. But in the countdown to Paris, many of these same governments signed or were pushing a raft of ambitious trade and investment deals that would pre-empt measures that they could take to deal with climate change (*see Box 1: Key mega deals being negotiated today on page 54*).

What we know of these deals so far, from the few texts that have leaked out of the secretive negotiations, is that they will lead to more production, more trade and more consumption of fossil fuels – at a time of global consensus on the need for reductions.[1] In particular, the EU-Canada Comprehensive Economic and Trade Agreement (CETA) and the EU-US Transatlantic Trade and Investment Partnership (TTIP) are expected to result in increased EU reliance on fossil fuel imports from North America, as well as a restriction of policy space to promote low carbon economies and renewables. The Trans-Pacific Partnership (TPP), a mega-pact involving 14 countries in Asia and the Americas that was concluded in February 2016, is expected to result in more gas exports from the US to the Pacific Rim countries. The new deals will also extend investor-state dispute settlement provisions, which companies are already using through the North American Free Trade Agreement (NAFTA) to reverse moratoriums on fracking and other popular environmental measures implemented by governments.[2]

Less has been said about how the provisions dealing with food and agriculture in these deals will affect our climate. But the question is vital, because food and farming figure hugely in climate change. From deforestation to fertilizer use, and from factory farms to supermarket shelves, producing, transporting, consuming and wasting food accounts for about half of all greenhouse gas emissions (GHGs).[3] Since creating new channels for the flow of farm goods and changing regulatory and investment regimes for agribusiness and the food industry are high priorities in the current deals, there will undoubtedly be impacts on climate change – and likely negative ones, unless we do something about it.

We see seven main ways through which the food and agriculture components of today's trade and investment deals will make the climate crisis worse.

1. Increasing production, trade and consumption of foods that are big emitters of greenhouse gases

Trade deals, on the face of it, are meant to increase trade. This includes trade in food.

The foods that make the biggest contribution to climate change are: red meat (worst: beef, lamb and pork), dairy (worst: butter and cheese, followed by milk and eggs), fish (worst: wild caught or industrially farmed), poultry, palm oil and highly processed foods (worst: those that are airfreighted). Of course, these are sweeping generalizations. There are numerous studies that try to measure the precise GHG emissions from different foods depending on where and how they are produced.[4] But roughly, the picture is what we see in Graph 1.

In terms of agricultural production, meat and dairy are the biggest contributors to climate change (*see Box 2: The elephant — er, lamb? — in the room on page 54*). Only 11 per cent of all meat produced is traded internationally, but globally speaking, meat production and consumption are projected to rise by 17 per cent by 2024 and outright double by 2050.[5] Increased trade

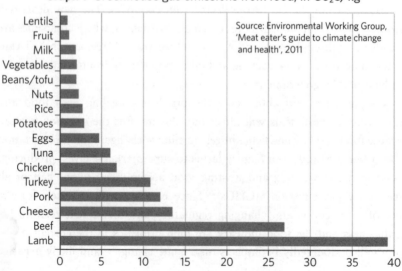

Graph 1 Greenhouse gas emissions from food, in CO_2e/kg

Source: Environmental Working Group, 'Meat eater's guide to climate change and health', 2011

is expected to play a role in that growth and some of this will come from the newest trade agreements, which could shift current meat trade dynamics quite a bit.[6] Of course, we cannot predict how much trade and consumption will grow as a direct result of these deals, but the tariff cuts and lower standards are expected to lead to increased supplies and therefore consumption in importing countries. That, after all, is what the industry lobbies are aiming for.

Take, for example, the TTIP. If it is signed, it is going to expand the European market for US beef, both high- and low-quality. Quotas for hormone-free beef will go up, while sanitary restrictions are going down.[7] European quality beef may not be able to compete, leading to a displacement of production to the US. Under CETA, Canada will be sending more pork, beef and dairy to Europe, while the EU will be exporting more cheese to Canada.

The recently concluded China-Australia free trade agreement (ChAFTA) is expected to play an important role in increased dairy production and trade in the Asia-Pacific region. China imports about 20 per cent of its dairy products and those imports are steadily rising.[8] Until now, because of the China-New Zealand trade deal, New Zealand dominated China's foreign dairy supply. Now Australia is expected to take some of that market. At the same time, Chinese companies themselves are investing heavily in offshore dairy production in Australia for export back to China.[9] They are also expanding their beef production base in New Zealand for export home.[10]

China's surging beef imports, which currently are permitted from just a handful of countries, grew by 18 per cent in the first half of 2015.[11] Australia now accounts for nearly half of that market because of ChAFTA.[12] Thanks to the China-New Zealand deal, China is the biggest buyer of New Zealand lamb and the second-biggest buyer of New Zealand beef.

Dairy trade was a very contentious issue in the TPP negotiations – one that reportedly held up the conclusion of the deal. Now that the deal has been concluded, Washington calls the US farm industry "the big winner" in the TPP, as not only US dairy exports are expected to grow significantly, but also US beef and pork.

Tariffs and quotas aside, markets are also expected to grow for certain agribusiness companies and their investors, due to the watering down of food safety regulations and labelling laws as a result of these new deals.[13] This is an important concern for farmers and consumers in quite a number of countries whose governments are negotiating. Unfortunately, despite statements from political leaders that nothing will change, many of the regulatory changes being pushed for by agribusiness giants involve lowering standards for chemicals, opening markets to cloned meat or genetically modified foods, and dropping disease-related barriers against poultry (avian flu) and beef (mad cow). Under the TPP, we now know that the US government secured the right to challenge other countries' food safety standards and to set new norms for the presence of genetically modified organisms in foods.[14] This will surely expand the US food industry's reach, globally.

2. Promoting industrial farming for export over local farms and food systems

Expansion of markets for European poultry and milk powder has long been a key facet of the EU's trade liberalization agendas, as African farmers and livestock keepers know. They have been mobilizing to stop the dumping of highly subsidized chicken and excess dairy from Europe for years. These struggles are now more and more connected to climate change. Industrial poultry, after all, is an important source of greenhouse gas emissions. Broilers, which are raised for their meat, produce seven times more GHG emissions than backyard birds. And layers, which are raised for their eggs, produce four times more.[15]

Chicken consumption is rising in many countries because it is a low-cost meat, and therefore global poultry trade is expected to increase. All of this trade comes from industrial poultry farms, which are higher emitting than backyard or small-scale operations. Brazilian and EU poultry farms are relatively high on the climate-unfriendliness scale, mostly attributed to their reliance on soybeans.[16] Even in China, where exports are just a small fraction of the country's production, trade deals are leading to increased imports of feed materials which serve the factory farms that are built with increased levels of foreign investment.

Beyond poultry, experts now say that over the next 10 years, increased global meat consumption will raise overall greenhouse gas emissions regardless of improved feed-to-meat conversion ratios in industrial production systems.[17]

3. Boosting global supermarkets and highly processed foods

The biggest names in food retail are aiming for growth in Asia, as well as in Africa and Latin America, through several of today's new trade agreements.

The expansion of global supermarkets brings with it the expansion of processed food production, trade and consumption. For example, under NAFTA, processed food consumption has skyrocketed in Mexico, bringing with it serious public health problems, and the country's retail sector has been taken over by large global chains.[18]

Processed foods – produced by Mondelēz, Nestlé, Pepsico, Danone, Unilever and the like – are important greenhouse gas emitters, not only because of all the energy used in packaging, processing and transporting the foods, but also because of the emissions generated on the farm. Processed foods are constructed out of the cheapest raw materials that companies can source from around the globe. One package of standard supermarket food can contain powdered milk from New Zealand, maize from the US, sugar from Brazil, soybeans from Argentina and palm oil from Indonesia – all foods that are high on the emissions scale.

One recent study of a box of Kellogg's breakfast cereal found that eating a 100-gram serving generates the equivalent of 264 grams of CO_2. Add milk to the cereal and the emissions go up by two to four times. The ingredients accounted for about half the total emissions from the cereal, while manufacturing, packaging and transport contributed the rest. The researchers identified more than 20 countries from which the ingredients were sourced, including maize from Argentina, milk powder from the EU, rice from Egypt and Thailand, wheat from Spain and sugar from the US.[19]

The growth of supermarkets and processed foods also means increased deforestation, and other changes in land and water use, to produce more sugar, maize, soybeans and palm oil – four products that form the backbone of the processed food sector. For example, in Nigeria, Wilmar, the largest palm oil trading company in the world, plans to expand its oil palm plantations in Cross River State and this, groups on the ground say, will inevitably mean new deforestation. Through its trade agreements with the Association of Southeast Asia Nations (ASEAN), India has become a major market for Indonesian and Malaysian palm oil, displacing coconut, mustard, groundnut, sesame and other traditional Indian vegetable oils, which were far less damaging to the climate. The same goes for China, the second-largest market for ASEAN palm oil after India.

The just-concluded TPP may bring an important upswing in palm oil production, trade and use. "I expect there to be quite a stampede of foreign investment in Southeast Asia when the final text of the agreement is published," Deborah Elms, the Executive Director of the Asian Trade Centre, told *The Wall Street Journal*.[20] Specifically, Malaysia's palm oil sector is supposed to attract a lot of this stampede, as investors jump in to lock down a new cheap source of oil for the US fast food industry.[21]

4. Climate cheating: the outsourcing of emissions

One of the effects of trade deals is that manufacturing is being outsourced to low-wage countries with few environmental restrictions. The countries where these products are consumed thus appear to have reduced emissions, when really those emissions have simply been transferred to the countries where the goods are now produced. As we see in the case of the US and China, neither country then wants to take responsibility. The same happens with foods.

Trade agreements favor food production in countries with low cost and/or heavily subsidized production, with high emissions levels. These countries have powerful industrial agriculture lobbies (US, Brazil, New Zealand, Europe) and are often heavily reliant on agriculture exports for their foreign revenues (US, Brazil, New Zealand, Ireland, Indonesia, Vietnam). It is highly unlikely that these countries will implement any measures to reduce emissions that might impinge on the competitiveness of their agricultural commodities. Already we see these countries moving with their companies to head off international efforts to make significant emissions cuts to agriculture, for instance with the Global Alliance for Climate Smart Agriculture.

The emissions imported with the foods are not likely to be accounted for by the importing country either. Even if an importing government were to try, measures to reduce imports of certain high greenhouse gas emitting commodities could be challenged as unfair trade restrictions under the new deals.

5. More biofuels

Biofuels are another form of polluting energy which, along with fossil fuels, may get a boost from the latest trade deals. This is especially when investment chapters of trade deals try to "level the playing field" for foreign investors by establishing rules on "national treatment" and "most favored nation", which makes access to land for the production of biofuels much easier. New patenting rules imposed through these deals also make it easier for corporations to engage in technology transfer, knowing that they will enjoy monopoly rights in the signatory countries. Already, EU climate policies have bolstered massive land grabbing in Africa for the production of ethanol for European markets. China, which currently sources ethanol from so-called free trade agreement partners Pakistan and Vietnam, is also investing heavily now in Brazil for this very purpose (a first-ever shipment of Brazilian ethanol for China just left South America). The Canadian biofuel industry expects to gain a new $50 million market opening in the EU thanks to CETA.[22] Many biofuel crops – sugar cane, sugar beet, sweet potato, oil palm, maize, sorghum, oilseed rape – can be interchangeably used in the food industry, too.

If the TTIP agreement between the US and the EU goes through, modellers say that the US will see a big increase in bioethanol and biodiesel production and exports to the EU which, conversely, will see a big rise in its sugar production and exports to the US.[23] The knock-on effects in Brazil, Argentina and China will be important, too.

Despite its poor scorecard in terms of human rights, land rights and carbon emissions, biofuel production is expected to be promoted increasingly as a renewable energy under climate mitigation strategies, and trade and investment deals will be facilitating this.

6. The promotion of local food economies undermined

"Buy national" or "buy local" programs, as well as country-of-origin labelling regulations, are generally considered discriminatory and trade distorting under so-called free trade doctrine. The World Trade Organization (WTO) did little to discourage these initiatives, but newfangled bilateral and regional trade deals could go much further. The EU particularly wants to gain much more access, for European companies, to US public markets at all levels (federal, state, local) under TTIP. Food sovereignty advocates and practitioners see this as a potential threat to local food economies that groups have been painstakingly building over the last decades (e.g. food policy council initiatives to support the use of local foods in public services like schools and hospitals).[24] Any moves to make "go local" or "use local" illegal in the food sector will automatically result in increased climate destabilization.[25]

The same is true of initiatives to support "green" purchasing or programs to require purchasing from small- and medium-sized enterprises in the name of mitigating climate change. Both of these types of effort can be contested by companies as discriminatory. Free trade agreements and investment treaties typically have an investor-state dispute mechanism that allows companies to challenge government policies like these. Sometimes the challenge results in huge financial compensation for the company on the losing end of such laws. Sometimes it causes governments to change policy to avoid such lawsuits.

Just like in the energy sector, we need to address consumption to address climate change. Increasing production and trade, or just making it greener, will not alleviate the problem. Since governments agree that 15 per cent of all global greenhouse gas emissions come from livestock and that 74 per cent of these come from beef and dairy, we have a great opportunity to positively eliminate a big part of the climate problem through local initiatives. But to do this, we need to defeat the trade deals and ideology that claim that promoting "local" economies is anti-free market and somehow bad for us. (It is only bad for the transnationals!)

7. Food security measures made illegal

In 2013, governments, prodded by corporate interests mainly coming from the US, tried to make it a WTO rule that public procurement of food stuffs in times of crisis should be considered a form of trade-distorting farm subsidy. Many governments purchase farm products from farmers to stabilize markets, provide guaranteed prices and run stockpiles or distribution systems in the public interest. The ravages caused by climate change (floods, drought, typhoons, for example) in a world of deregulation and corporation concentration make food shocks more common and more threatening. That means these basic food security measures and strong public procurement programs are more and more needed. Ironically, as soon as the Paris climate talks ended in December 2015, governments flew to Nairobi for a WTO ministerial meeting to decide whether such measures will be considered lawful or not under the global trade regime.

Time to stop destabilizing the climate!

Food consumption patterns are shifting. The Western diet is spreading, particularly in the Global South, bringing with it problems of health, but also increasing climate pressure. Some people say we need diet change, not climate change. Commodity traders, agribusiness firms, retail chains, private equity groups and other kinds of corporations that finance and run the industrial food system have a keen interest in expanding business in those very markets. Trade agreements are a great tool to do that, but it's not just a North-South affair. Brazilian companies are competing with Thai counterparts for emerging market shares in Africa, Russia or the Middle East. Australia wants a bigger part of the action in China, which is doing more business with the US, and so on.

We have to wake up and do the math. If we want to deal with climate change, we have to cut consumption of some foods and that means cutting production and trade as well. Luckily, it is quite do-able. But it does require a structural scaling back of "Big Food" and "Big Retail" and those who finance them. Instead, small- and medium-sized farms, processing and markets supported by public procurement and financing, could do the job better. It requires a push, and bringing the different struggles around climate change together with the struggles for food sovereignty and against corporate-driven trade agreements.

What to do?

Join the growing campaigns against major trade deals like TTIP, TPP, RCEP, TiSA and CETA. See www.bilaterals.org for links to key groups and more information.

- Start a focused campaign on trade, climate and food to show how trade deals your government is negotiating will specifically affect greenhouse gas emissions from food and get them stopped.
- Raise the issue of food and food trade in local discussions and actions you're involved in to battle climate change.
- Use your imagination to develop concrete initiatives to reduce (y) our reliance on the industrial food system and shrink demand for their products. Start a boycott action – this is what food industry leaders fear most.
- Become more aware about the climate impact of the foods you eat and initiate, join or strengthen a local food initiative, be it a co-op, school program, an AMAP (Association for the maintenance of peasant agriculture), a CSA (Community-supported agriculture scheme), or a farmers' market.

Box 1: Key mega deals being negotiated today

CETA: Comprehensive Economic and Trade Agreement between the EU and Canada. The negotiations were completed in 2014, but the text still needs to be ratified. There is talk of still tweaking some of the language on investor protection, given the scale of public outcry about it.

FTAAP: Free Trade Area of Asia and the Pacific, a trade pact that aims to cover all member states of the Asia Pacific Economic Co-operation (APEC). Was originally floated by the US but now is championed by China as a counterweight to the TPP (which excludes China). Negotiations have not yet begun.

TiSA: Trade in Service Agreement, a very significant pact being secretly negotiated among 40 countries outside the World Trade Organization. Aims to set new global standards for trade in services for all future trade deals.

TTIP or **TAFTA**: Transatlantic Trade and Investment Partnership between the EU and the US. Is under negotiation, but massively contested by civil society.

TPP or **TPPA**: Trans-Pacific Partnership, recently concluded among 14 countries on both sides of the Pacific (Australia, Brunei, Canada, Chile, Japan, Malaysia, Mexico, New Zealand, Peru, Singapore, US, Vietnam). Will need to be ratified by national parliaments.

RCEP: Regional Comprehensive Economic Partnership is a trade agreement between the 10 member Association of Southeast Asia Nations (Brunei, Burma, Cambodia, Indonesia, Laos PDR, Malaysia, Philippines, Singapore, Thailand, Vietnam) and six neighbors: Australia, China, India, Japan, New Zealand and South Korea. Currently being negotiated behind closed doors.

Box 2: The elephant – er, lamb? – in the room

The meat industry is perhaps the biggest single cause of climate change. The data vary, are debated and may be distorted. For example, there is a tendency in some corners to present super-industrialized cattle operations in the US or Western Europe as being more "climate friendly" than sustainable grazing systems in India or Niger. That is because agencies like the FAO tend to use a narrow lens of "efficiency" to make the comparison and they don't factor in the positive climate contributions from sustainable grazing systems in Asia or Africa. Even the IPCC, which produces much of the "science" that people rely on to judge and act on climate change, gets it wrong sometimes. Still, there is no reason to doubt that raising or capturing animals for food is one of the biggest causes of climate change.

Some key facts worth chewing on:
- According to one often cited but highly criticized study by the FAO in 2006, livestock are responsible for 18 per cent of all greenhouse gas emissions. Researchers from the World Bank, writing for the Worldwatch Institute in 2009, put it at 51 per cent. In 2013, the FAO reduced its figure to 15 per cent. Either way, it's big – more than all forms of transportation (air, car, ship) combined.
- Two-thirds (65%) of livestock emissions comes from beef (35%) and dairy (30%) production alone, FAO reported in 2013.[26] World dairy production is responsible for four per cent of all global GHG emissions.
- One quarter of the earth's land mass is used for grazing and nearly half of all crops that we produce (40%) – which produce GHGs as well – are fed to livestock.
- Livestock contribute to climate change not so much in terms of carbon emissions, but in terms of methane (from ruminant digestion systems = 47% of their emissions) and nitrous oxide (from the fertilizer used to produce their feed + animal waste = 24% of livestock emissions). Methane and nitrous oxide are far more dangerous for our climate than carbon dioxide. In fact, recent data from the University of Minnesota, Yale and USDA suggests that the IPCC has been underestimating N_2O emissions from industrial crop production – much of this to produce animal feed – by 40 per cent.

Take into account the general thinking that the world's meat and dairy consumption are projected to double by 2050, and one can see this is a serious and growing problem.

The good news is that we can do something about this, and relatively quickly. Cutting back on meat and dairy production, consumption and trade would be an effective and realistic way to reduce climate chaos. Compared to carbon, methane is a lot easier and a lot faster to "clean up" from the atmosphere. As to nitrous oxide, a contraction and restructuring of the meat industry towards small-scale and local systems could do away with a lot of the fertilizer that is currently being used to produce feed.

We don't have to all go vegan, but if we want to address climate change we have to take some very serious action towards the meat industry on a systemic and international scale. It's not enough to stop extracting and burning fossil fuels.

(It's important to note that FAO data on GHG emissions from livestock is produced with input from people from the meat and dairy industry: the International Poultry Council, International Feed Industry Federation, International Meat Secretariat, International Egg Commission and Danone.)

The original fully referenced version of this article can be found at:
https://www.grain.o1rg/e/5317

1 See forthcoming reports from Corporate Europe Observatory, http://corporateeurope.org, as well as previous reports from Sierra Club, the Friends of the Earth network, CEO and others compiled at http://www.bilaterals.org/?+-climate-+

2 Rossman, P. (2015) "Against the Trans-Pacific Partnership", *Jacobin*, 13 May, https://www.jacobinmag.com/2015/05/trans-pacific-partnership-obama-fast-track-nafta/

3 See La Via Campesina and GRAIN (2014) "Food sovereignty: 5 steps to cool the planet and feed its people", 5 December, https://www.grain.org/e/5102

4 We are not in a position to assess that data here, but hope to do so soon.

5 See OECD-FAO (2015) "Agricultural Outlook 2015", 1 July, http://dx.doi.org/10.1787/agr_outlook-2015-10-en. Seafood trade has already doubled in the last five years and become the most widely traded protein. For more info, see Rabobank (2015), "Seafood: A myriad of globally traded aquatic products", 24 March, http://rabobank-food-agribusiness-research.pr.co/98495-seafood-a-myriad-of-globally-traded-aquatic-products

6 See the "expanded" meat chapter in OECD-FAO, *op cit.*

7 The allowed tonnage for hormone-free beef might be raised by perhaps 50,000 tonnes per year. This is a hypothetical that analysts are working with, reflecting what the EU offered

Canada under CETA. See http://capreform.eu/ttip-and-the-potential-for-us-beef-imports/

8   Gannon, G. and Smith, S. (2015) "China FTA: Australian dairy to win share from New Zealand", *Weekly Times*, 26 May, http://www.weeklytimesnow.com.au/agribusiness/dairy/china-fta-australian-dairy-to-win-share-from-new-zealand/story-fnkeqg0i-1227369585925. Also see: "China dairy sector", CLAL.it, http://www.clal.it/en/?section=stat_cina

9   Chinese investors are not the largest foreign landholders in Australia, but are buying or bidding for some of the country's most significant cattle and dairy farm operations. See www.farmlandgrab.org

10 See for example, Tajitsu, N. and Greenfield, C. (2015) "China's Bright to buy 50 pct stake in NZ meat processor", *Reuters*, 14 September, http://www.reuters.com/article/2015/09/15/newzealand-silverfern-merger-idUSL4N11L1E820150915

11 Dimsums (2015) "China's agricultural imports in disarray", 15 August, http://dimsums.blogspot.fr/2015/08/chinas-agricultural-imports-in-disarray.html

12 China Daily (2015) "Pengxin may buy two cattle farms in Australia", August 29, http://www.ecns.cn/business/2015/08-29/179146.shtml

13 See GRAIN (2013) "Food safety in the EU-US trade agreement: going outside the box", 10 December, https://www.grain.org/e/4846. Also see: FoEE, GRAIN et al (2015) "EU-US trade deal threatens food safety", 5 February, https://www.grain.org/e/5129

14 Weaver, M. (2015) "Vilsack: TPP text available in next 30 days", *Capital Press*, 6 October, http://www.capitalpress.com/Nation_World/Nation/20151006/vilsack-tpp-text-available-in-next-30-days

15 Data are from FAO Global Livestock Environmental Assessment (GLEAM) report (2013), "Greenhouse gas emissions from pig and chicken supply chains", http://www.fao.org/docrep/018/i3460e/i3460e.pdf

16 *Idem.*, Figure 36, p. 55

17 *Idem.*

18 See GRAIN (2015) "Free trade and Mexico's junk food epidemic", 2 March https://www.grain.org/e/5170

19 Jeswani, H.K. et al (2015) "Environmental sustainability issues in the food-energy-water nexus: Breakfast cereals and snacks", *Science Direct*, April http://www.sciencedirect.com/science/article/pii/S2352550915000238

20 Watts, J.M. et al (2015) "Company stampede to Southeast Asia seen on Trans-Pacific Partnership trade pact," *Wall Street Journal*, 7 October, http://www.wsj.com/articles/company-stampede-to-southeast-asia-seen-on-trade-pact-1444230531

21 Bernama (2015) "TPP broadens market scope in US, say palm oil experts", Malaymail Online, 7 October, http://www.themalaymailonline.com/money/article/tpp-broadens-market-scope-in-us-say-palm-oil-experts

22 Government of Canada, "CETA: What has been said", http://www.international.gc.ca/trade-agreements-accords-commerciaux/agr-acc/ceta-aecg/benefits-avantages/quotes-citations.aspx

23 Beghin, J. et al (2014) "The impact of an EU-US Transatlantic Trade and Investment Partnership agreement on biofuel and feedstock markets", J Working Paper 14-WP 552, November, http://www.card.iastate.edu/publications/dbs/pdffiles/14wp552.pdf

24 See Hansen-Kuhn, K. (2014) "Local economies on the table: TTIP procurement update", IATP, 13 November, http://www.iatp.org/documents/local-economies-on-the-table

25 Not all "go local" initiatives in the food sector are better for the climate. But a lot are.

26 FAO (2013) "Major cuts of greenhouse gas emissions from livestock within reach, Key facts and findings", 26 September, http://www.fao.org/news/story/en/item/197623/icode/

2
Hungry for land

2.1 The solution to climate change is in our lands

A global effort to give small farmers and indigenous communities control over lands is the best hope we have to deal with climate change and to feed the world's growing population.

As governments converged on Lima for the December 2014 UN Climate Change Conference, the brutal killing of Peruvian indigenous activist Edwin Chota and three other Ashaninka men in September of the same year shone a spotlight on the connection between deforestation and indigenous land rights. The simple truth is plain to see: the most effective and just way to prevent deforestation and its impacts on the climate is to recognize and respect the sovereignty of indigenous peoples over their territories.

Peru's violent land conflicts also bring into focus another issue of equal importance to climate change that can no longer be ignored: the concentration of farmland in the hands of a few.

Small farms of less than five hectares represent 78 per cent of all farms in Peru, but occupy a mere six per cent of the country's agricultural lands. This disturbing figure mirrors the global situation. Worldwide, small farms account for 90 per cent of all farms, yet they occupy less than one quarter of the agricultural land. This is bad news for the climate.

Just as the dispossession of indigenous peoples of their territories has opened the door to destructive, unsustainable resource extraction, the dispossession of peasants of their lands has laid the basis for an industrial food system that, among its many negative effects, is responsible for between 44 per cent and 57 per cent of all global greenhouse gas emissions.

Food does not have to make such an overweight contribution to climate change. GRAIN estimates that a worldwide redistribution of lands to small farmers and indigenous communities, combined with policies to support local markets and cut the use of chemicals, can reduce global greenhouse gas emissions by half within a few decades and significantly curb deforestation. Simply by rebuilding the organic matter that has been depleted by decades of industrial agriculture, small farmers can put one quarter of the excess carbon dioxide that is now in the atmosphere back into the soil.

Giving lands back to small farmers and indigenous communities is also the most effective way to deal with the challenges of feeding a growing global population in an era of climate chaos. The available global data show that small farmers are more efficient than big plantations at producing food. On the fraction of lands that they have held on to, small farmers and indigenous communities continue to produce most of the world's food – 80 per cent of the food in developing countries, says the FAO. Even in Brazil, a powerhouse of industrial agriculture, small farms occupy one quarter of the farmlands but produce 87 per cent of the country's cassava, 69 per cent of its beans, 59 per cent of its pork, 58 per cent of its cow's milk, 50 per cent of its chickens, 46 per cent of its maize, 33.8 per cent of its rice and 30 per cent of its cattle.

The twin needs of feeding the world and cooling the planet can be met. But not if the governments continue to ignore and violently repress the struggles of their peasants and indigenous peoples for land.

Original, fully referenced article by GRAIN and Via Campesina can be found at https://www.grain.org/e/5105

2.2 Family farm stories are not the fairy tales we're being told

The United Nations declared 2014 as the International Year of Family Farming.[1] As part of the celebrations, the UN Food and Agriculture Organization (FAO) released its annual "State of Food and Agriculture",[2] which in 2014 was dedicated to family farming. Family farmers, FAO says, manage between 70 per cent and 80 per cent of the world's farmland and produce 80 per cent of the world's food.

But on the ground – whether in Kenya, Brazil, China or Spain – rural people are being marginalized and threatened, displaced, beaten and even killed by a variety of powerful actors who want their land.

A recent comprehensive survey[3] by GRAIN, examining data from around the world, finds that although small farmers feed the world, they are doing so with just 24 per cent of the world's farmland – or 17 per cent if you leave out China and India. GRAIN's report also shows that this meagre share is shrinking fast.

How, then, can FAO claim that family farms occupy 70 per cent to 80 per cent of the world's farmland? In the same report, FAO claims that only one per cent of all farms in the world are larger than 50 hectares, and that these few farms control 65 per cent of the world's farmland, a figure much more in line with GRAIN's findings.

The confusion stems from the way FAO deals with the concept of family farming, which it roughly defines as any farm managed by an individual or a household. (They admit there is no precise definition. Various countries, like Mali, have their own.) Thus, a huge industrial soya bean farm in rural Argentina, whose family owners live in Buenos Aires, is included in FAO's count of "family farms". What about sprawling Hacienda Luisita, owned by the powerful Cojuanco family in the Philippines and for decades the epicenter of the country's battle for agrarian reform. Is that a family farm?

Looking at ownership to determine what is and is not a family farm masks all the inequities, injustices and struggles that peasants and other small-scale food producers across the world are mired in. It allows FAO to paint a rosy picture and conveniently ignore perhaps the most crucial factor affecting the capacity of small farmers to produce food: lack of access to land. Instead, the

FAO focuses its message on how family farmers should innovate and be more productive.

Small food producers' access to land is shrinking due to a range of forces. One is that because of population pressure, farms are getting divided up among family members. Another is the vertiginous expansion of monoculture plantations. During the last 50 years, a staggering 140 million hectares – the size of almost all the farmland in India – has been taken over by four industrial crops: soya bean, oil palm, rapeseed and sugar cane. And this trend is accelerating.

In the next few decades, experts predict that the global area planted to oil palm will double,[4] while the soybean area will grow by one third.[5] These crops don't feed people. They are grown to feed the agro-industrial complex.

Other pressures pushing small food producers off their land include the runaway plague of large-scale land grabs by corporate interests. In the last few years alone, according to the World Bank, some 60 million hectares of fertile farmland have been leased on a long-term basis to foreign investors and local elites, mostly in the Global South. Although some of this is for energy production, a big part of it is to produce food commodities for the global market, instead of family farming.

Small works better

The paradox, however, and one of the reasons why despite having so little land, small producers are feeding the planet, is that small farms are often more productive than large ones.

If the yields achieved by Kenya's small farmers were matched by the country's large-scale operations, its agricultural output would double. In Central America, the region's food production would triple. If Russia's big farms were as productive as its small ones, output would increase by a factor of six.

Another reason why small farms are feeding the planet is because they prioritize food production. They tend to focus on local and national markets and their own families. In fact, much of what they produce doesn't enter into trade statistics – but it does reach those who need it most: the rural and urban poor.

If the current processes of land concentration continue, then no matter how hard-working, efficient and productive they are, small farmers will simply not be able to carry on. The data show that the concentration of farmland in fewer and fewer hands is directly related to the increasing number of people going hungry every day.

According to one UN study,[6] active policies supporting small producers and agro-ecological farming methods could double global food production in a decade and enable small farmers to continue to produce and utilize bio-diversity, maintain ecosystems and local economies, while multiplying and strengthening meaningful work opportunities and social cohesion in rural areas. Agrarian reforms can and should be the springboard to moving in this direction.

Experts and development agencies are constantly saying that we need to double food production in the coming decades. To achieve that, they usually recommend a combination of trade and investment liberalization plus new technologies. But this will only empower corporate interests and create more inequality. The real solution is to turn control and resources over to small producers themselves and enact agricultural policies to support them.

The message is clear. We urgently need to put land back in the hands of small farmers and make the struggle for genuine and comprehensive agrarian reform central to the fight for better food systems worldwide.

The FAO's lip service to family farming just confuses the matter and avoids putting the real issues on the table.

Original article: https://www.grain.org/e/5072

1 FAO (2014) http://www.fao.org/family-farming-2014/en/
2 FAO (2014) http://www.fao.org/publications/sofa/2014/en/
3 GRAIN (2014) "Hungry for Land: small farmers feed the world with less than one-quarter of all farmland", http://www.fao.org/publications/sofa/2014/en/
4 See Corley, R.H.V. (2009) "How much palm oil do we need?", *Environmental Science and Policy* 12(2): 134–9, http://www.sciencedirect.com/science/article/pii/S1462901108001196
5 Alexandratos, N. and Bruinsma, J. (2012) "World Agriculture towards 2030/2050, The 2012 Revision", ESA Working Paper 12-03, http://www.fao.org/docrep/016/ap106e/ap106e.pdf
6 GRAIN (2011) "Eco-farming can double food production in 10 years", 8 March, https://www.grain.org/bulletin_board/entries/4219-eco-farming-can-double-food-production-in-10-years

2.3 Hungry for land: small farmers feed the world with less than one quarter of all farmland

It is commonly heard today that small farmers produce most of the world's food. But how many of us realize that they are doing this with less than one quarter of the world's farmland, and that even this meagre share is shrinking fast? If small farmers continue to lose the very basis of their existence, the world will lose its capacity to feed itself.

GRAIN took an in depth look at the data to see what is going on and the message is crystal clear. We urgently need to put land back in the hands of small farmers and make the struggle for agrarian reform central to the fight for better food systems.

Governments and international agencies frequently boast that small farmers control the largest share of the world's agricultural land. Inaugurating 2014 as the International Year of Family Farming, José Graziano da Silva, the Director General of the United Nations Food and Agriculture Organization (FAO), sang the praises of family farmers but didn't once mention the need for land reform. Instead he stated that family farms already manage most of the world's farmland[1] – a whopping 70 per cent, according to his team.[2,3,4] Another report published by various UN agencies in 2008 concluded that small farms occupy 60 per cent of all arable land worldwide.[5] Other studies have come to similar conclusions.[6]

But if most of the world's farmland is in small farmers' hands, then why are so many of their organizations clamoring for land redistribution and agrarian reform? Because rural peoples' access to land is under attack everywhere. From Honduras to Kenya and from Palestine to the Philippines, people are being dislodged from their farms and villages. Those who resist are being jailed or killed. Land is becoming more and more concentrated in the hands of the rich and powerful.

For rural people, land and territories are the backbone of their identities, their cultural landscape and their source of wellbeing. Yet land is being taken away from them and concentrated in fewer and fewer hands at an alarming pace. Then, there is the other part of the picture: that concerning food.

Although it is now increasingly common to hear that small farmers produce the majority of the world's food, we are also constantly being fed the message that a "more efficient" industrial food system is needed to feed the world. At the same time, we are told that 80 per cent of the world's hungry people live in rural areas, many of them farmers or landless farmworkers.

How do we make sense of all this? GRAIN decided to take a closer look at the facts.[7,8]

The figures and what they tell us

When we looked at the data, we came across quite a number of difficulties. Countries define "small farmer" differently. There are no centralized statistics on who has what land. There are no databases recording how much food comes from where. And different sources give widely varying figures for the amount of agricultural land available in each country.

In compiling the figures, we used official statistics from national agricultural census bureaus in each country wherever possible, complemented by FAOSTAT (FAO's statistical database) and other FAO sources where necessary. For statistical guidance on what a small farm is, we generally used the definition provided by each national authority, since the conditions of small farms in different countries and regions can vary widely. Where national definitions were not available, we used the World Bank's criteria.

In light of this, there are important limitations to the data – and our compilation and assessment of them. The dataset that we produced is fully referenced and publicly available online.[9]

Despite the shortcomings of the data, we feel confident in drawing six major conclusions:

1. The vast majority of farms in the world today are small and getting smaller.
2. Small farms are currently squeezed onto less than one quarter of the world's farmland.
3. We are fast losing farms and farmers in many places, while big farms are getting bigger.
4. Small farms continue to be the major food producers in the world.
5. Small farms are overall more productive than big farms.
6. Most small farmers are women.

Two things shocked us.

One was to see the extent of land concentration today. What we see happening in many countries is a kind of reverse agrarian reform, whether it's through corporate land grabbing in Africa, the massive expansion of soybean plantations in Latin America, or the extension of the European Union and its agricultural model eastward. Control over land is being usurped from small producers and their families, with elites and corporate powers pushing people onto smaller and smaller land holdings, or off the land entirely into camps or cities.

The other shock was to find that, today, small farms have less than one quarter of the world's agricultural land – or less than one fifth if one excludes China and India from the calculation. Such farms are getting smaller all the time, and if this trend persists they might not be able to continue to feed the world.

Let's go through these findings point by point.

1. The vast majority of farms in the world today are small and getting smaller
By our calculations, more than 90 per cent of all farms worldwide are small, holding on average 2.2 hectares (*See Table 1 on page 68*). Even if we exclude China and India – where about half of the world's small farms are located – small farms still account for more than 85 per cent of all farms on the planet.

Due to myriad forces and factors (such as land concentration, population pressure or lack of access to land) most small farms have been getting smaller over time. Average farm sizes have shrunk in Asia and Africa. In India, the average farm size roughly halved from 1971 to 2006, doubling the number of farms measuring less than two hectares. In China, the average area of land cultivated per household fell by 25 per cent between 1985 and 2000, after which it slowly started to increase due to land concentration and industrialization. In Africa, average farm size is also falling.[10]

2. Small farms are being squeezed onto less than one quarter of global agricultural land
Table 1 reveals another stark fact: globally, small farms have less than 25 per cent of the world's farmland today. If we exclude India and China again, then the reality is that small farms control less than one fifth of the world's farmland: 17.2 per cent, to be precise.

Table 1: Global distribution of agricultural land

	Agricultural land (hectares)	Number of farms	Number of small farms	Small farms as % of all farms
Asia-Pacific	1,990,228	447,614	420,348	93.9%
China	521,775	200,555	200,160	99.8%
India	179,759	1 38,348	127,605	92.2%
Africa	1,242,624	94,591	84,757	89.6%
Latin America & Caribbean	894,314	22,333	17,894	80.1%
North America	478,436	2,410	1,850	76.8%
Europe	474,552	42,013	37,182	88.5%
TOTAL	**5,080,154**	**608,962**	**562,031**	**92.3%**

	Agricultural land in the hands of small farmers (thousands of ha)	% of agricultural land in the hands of small farmers	Average size of small farms (ha)
Asia-Pacific	689,737	34.7%	1.6
China	370,000	70.9%	1.8
India	71,152	39.6%	0.6
Africa	182,766	14.7%	2.2
Latin America & Caribbean	172,686	19.3%	9.7
North America	125,102	26.1%	67.6
Europe	82,337	17.4%	2.2
TOTAL	**1,252,628**	**24.7%**	**2.2**

Notes: All figures on agricultural land obtained from FAOSTAT
(http://faostat3.fao.org/faostat-gateway/g.o/to/home/E).
All figures on number and size of farms obtained from national authorities, as far as possible
(see regional tables for details).

India and China merit special attention because of the huge number of farms and farmers they are home to. In these two countries, small farms still occupy a relatively large percentage of farmland.

We find the most extreme disparities in 30 of the countries for which we have sufficient data. Here, more than 70 per cent of farms are small, but they are relegated to less than 10 per cent of the country's farmland. These worst cases are listed in Table 2.

Table 2: Worst-case scenarios

Countries where more than 70% of farms are small, yet control less than 10% of domestic agricultural land	
Africa	Algeria, Angola, Botswana, Congo, DR Congo, Guinea, Guinea-Bissau, Lesotho, Madagascar, Mali, Morocco, Mozambique, Namibia, Zambia
Americas	Chile, Guyana, Panama, Paraguay, Peru, Venezuela
Asia	Iran, Jordan, Kyrgyzstan, Lebanon, Malaysia, New Zealand, Qatar, Turkmenistan, Yemen
Europe	Bulgaria, Czech Republic, Russia

3. We're fast losing farms and farmers in many places, yet big farms are getting bigger

Almost everywhere, big farms have been accumulating more land in recent decades, with many small and medium-sized farmers going out of business as a result.

The situation seems most dramatic in Europe, where decades of EU agricultural policies have led to the loss of millions of farms. In Eastern Europe, the process of land concentration started earnestly after the fall of the Berlin Wall and the enlargement of the European Union. Millions of farmers were forced out of business by the opening up of East European markets to subsidized farm produce from the West. In Western Europe, meanwhile, biased agricultural policies, coupled with large-scale infrastructure, transportation and urbanization projects, have taken a vicious toll. Large farms now represent less than one per cent of all farms in the European Union as a whole, but control 20 per cent of EU farmland.[11,12,13]

Official data on farm losses and land concentration in Africa and Asia are harder to get, and the situation there is less clear, since contradictory factors and forces are often at play. In many countries with high levels of population growth, the number of small farms actually increases as small farms are divided up between children. But at the same time, land concentration is growing.

The rapid expansion of huge industrial commodity farms is a relatively recent phenomenon in Africa, although it has been going on for decades in many countries of Latin America and in several parts of Asia. The conclusion is inescapable: across the world more and more fertile agricultural land is occupied by huge farms to produce industrial commodities for export, pushing small producers into an ever-decreasing share of the world's farmland.

Box 1: The invasion of the mega-farms

Why are small farmers increasingly pushed into an ever-smaller corner of the world's farmland? There are many complex factors and forces at play. One is population growth in rural areas in many countries, where small farmers are increasingly forced to divide their land among their children, resulting in smaller and smaller farms, as they have no access to more land. Another is urbanization and the covering of fertile farmland with concrete to serve expanding cities and their transportation needs. The spread of extractive industries (mining, oil, gas and now fracking), tourism and infrastructure projects are other contributors.

Perhaps the single most important factor is the tremendous expansion of industrial commodity crop farms. The food and energy industries are shifting farmland and water away from local food production to the production of commodities for industrial processing. Since the 1960s, a massive 140 million hectares of fields and forests have been taken over to grow just four crops – soybean, oil palm, rapeseed and sugar cane.

To put things in perspective: this is roughly the same area as all the farmland in the European Union. And the invasion is clearly accelerating: almost 60 per cent of this land-use change occurred in the last two decades. This doesn't take into account any of the other crops that are fast becoming industrial commodities produced on mega-farms or the tremendous growth of the industrial forestry sector. The FAO calculates

---

that in developing countries alone, monoculture tree plantations grew by more than 60 per cent, from 95 to 154 million ha, just between 1990 and 2010. Others put this figure higher, and point out that the trend is accelerating.[14] Many of these new plantations are encroaching on natural forests, but they are also increasingly taking over farmland.

Without significant changes in government policies, this aggressive attack by commodity monocultures is set to expand further. According to the FAO, between now and 2050 the world's soybean area is set to increase by one third to some 125 million hectares, the sugar cane area by 28 per cent to 27 million hectares, and the rapeseed area by 16 per cent to 36 million hectares.[15] As for oil palm, there are currently 15 million hectares under production for edible palm oil (not biofuels), and this is expected to nearly double, with an additional 12–29 million hectares coming into production by 2050.[16] Much of this expansion will happen in Africa, Asia and Latin America. Soybean and sugar cane are today mostly produced in Latin America, and oil palm in Asia, but these crops are also now being pushed aggressively into Africa as part of the global wave of land grabbing.

Graph 1 The global encroachment of the industrial crops

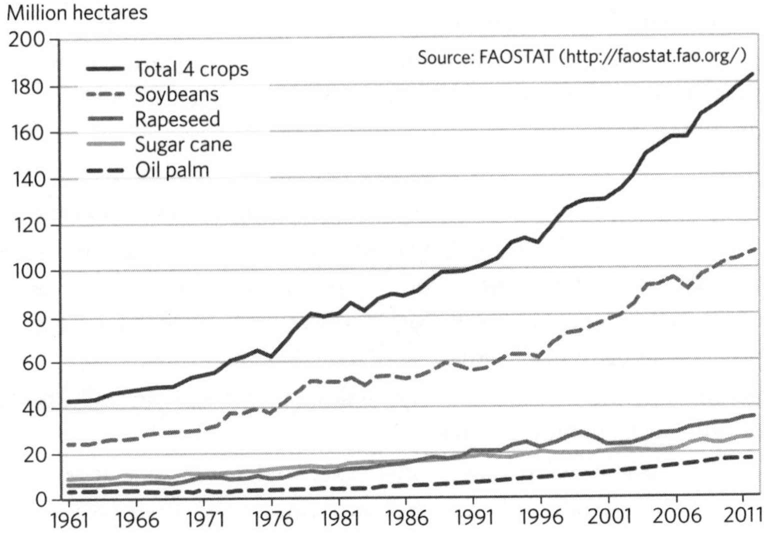

Million hectares

Source: FAOSTAT (http://faostat.fao.org/)

Legend:
— Total 4 crops
--- Soybeans
— Rapeseed
— Sugar cane
-- Oil palm

This trend is compounded by another recent phenomenon: the new wave of land grabbing. The World Bank has estimated that between 2008 and 2010 at least 60 million hectares of fertile farmland were leased out or sold to foreign investors for large-scale agricultural projects, with more than half of this occurring in Africa.[17] These massive new agribusiness projects are throwing an incalculable number of small farmers, herders and indigenous people off their territories.[18] Yet no-one seems to have a real grasp of how much land has changed hands through these deals. The scores, possibly hundreds, of millions of hectares of agricultural land being taken away from rural communities are not yet captured in the official statistics that were available for this report.

4. Despite their scarce and dwindling resources, small farmers are the world's major food producers

At a time when agriculture is almost exclusively judged in terms of its capacity to produce commodities, we tend to forget that the main role of farming is feeding people. This bias has infiltrated national census data, too, as many nations do not include questions about who produces what and with what means. However, when that information is available, a clear picture emerges: small farmers still produce most of the food. The UN Environment Programme, the International Fund for Agricultural Development, FAO and the UN Special Rapporteur on the Right to Food all estimate that small farmers produce up to 80 per cent of food in the non-industrialized countries.[19] Across a diverse range of countries, the data show that small farmers produce a much larger proportion of their nations' food than might be expected from their limited landholdings:

- Brazil: 84 per cent of farms are small and control 24 per cent of the land, yet they produce: 87 per cent of cassava; 69 per cent of beans; 67 per cent of goat's milk; 59 per cent of pork; 58 per cent of cow's milk; 50 per cent of chickens; 46 per cent of maize; 38 per cent of coffee; 33.8 per cent of rice; and 30 per cent of cattle[20]
- Kenya: With just 37 per cent of the land, small farms produced 73 per cent of agricultural output in 2004[21]
- Russia: Small farms have 8.8 per cent of the land, but provide 56 per cent of agricultural output, including: 90 per cent of potatoes; 83 per cent of vegetables; 55 per cent of milk; 39 per cent of meat; and 22 per cent of cereals[22]

If small farmers have so little land, how can they provide most of the food in so many countries? One reason is that small farms tend to be more productive than big ones, as we explain in the next section. But another factor is this historical constant: small or peasant farms prioritize food production. They tend to focus on local and national markets and their own families. Much of what they produce doesn't enter into national trade statistics, but it does reach those who need it most: the rural and urban poor.

Big corporate farms, on the other hand, tend to produce commodities and concentrate on export crops, many of which people can't eat as such. These include plants grown for animal feed or biofuels, wood products and other non-food crops. The primary concern for corporate farms is their return on investment, which is maximized at low levels of spending and thus often implies less intensive use of the land. The expansion of giant monoculture plantations, as discussed earlier, is part of this picture. Large corporate farms also often have considerable reserves of land that lie unused until land that is currently being cropped or grazed is exhausted. International development agencies are constantly warning that we need to double food production in the coming decades. To achieve that, they usually recommend a combination of trade and investment liberalization plus new technologies. But this will only create more inequality. The real solution is to turn control and resources over to small producers themselves and enact agricultural policies to support them.

In a recent paper on small farmers and agro-ecology, the UN Special Rapporteur on the Right to Food concludes that global food production could be doubled within a decade if the right policies towards small farmers and traditional farming were adopted. Reviewing the currently available scientific research, he shows that agro-ecological initiatives by small farmers themselves have already produced an average crop yield increase of 80 per cent in 57 developing countries, with an average increase of 116 per cent among all African initiatives assessed. Recent projects conducted in 20 African countries provided a doubling of crop yields in a short period of just 3 to 10 years.[23,24,25,26]

The real question, then, is how much more food could be produced if small farmers had access to more land and could work in a supportive policy environment rather than under the siege conditions they are facing today?

5. Small farms not only produce most of the food, they are also the most productive

For some, the idea that small farms are more productive than big farms might seem counterintuitive. After all, we have been told for decades that industrial farming is more efficient and more productive. It's actually the other way around. The inverse relationship between farm size and productivity has been long established and is dubbed "the productivity paradox".[27]

In the EU, 20 countries register a higher rate of production per hectare on small farms than on large farms. In nine EU countries, productivity of small farms is at least twice that of big farms.[28] In the seven countries where large farms have higher productivity, it is only slightly higher than that of small farms.[29] This tendency is confirmed by numerous studies in other countries and regions, all of them showing higher productivity on small farms. Our data indicate, for example, that if all farms in Kenya had the current productivity of its small farms, Kenya's agricultural production would double. In Central America and Ukraine, it would almost triple. In Hungary and Tajikistan it would increase by 30 per cent. In Russia, it would be increased by a factor of six.[30]

Although big farms generally consume more resources, control the best lands, receive most of the irrigation water and infrastructure, get most of the financial credit and technical assistance, and are the ones for which most modern inputs are designed, they have lower technical efficiency and therefore lower overall productivity. Much of this has to do with low levels of employment used on big farms in order to maximize return on investment.[31] Beyond strict productivity measurements, small farms also are much better at producing and utilizing biodiversity, maintaining landscapes, contributing to local economies, providing work opportunities and promoting social cohesion, not to mention their real and potential contribution to reversing the climate crisis.[32]

6. Most small farmers are women, but their contributions are ignored and marginalized

The role of women in feeding the world is not adequately captured by official data and statistical tools. The FAO, for example, defines women who are "economically active in agriculture" as only those who get a monetary income from farming. Using this concept, FAOSTAT indicates that 28 per cent of the rural population in Central America are "economically active" and that women form just 12 per cent of that group![33] This distorted view does not change significantly from country to country. However, when data is more specific, a totally different picture emerges. The last published agricultural census figures from El Salvador indicate that women are just 13 per cent of "producers", meaning farm holders, much in line with the number provided by the FAO.[34] However, the same census indicates that women provide 62 per cent of the labor force used on family farms. The situation in Europe is better for women, but still highly unequal. There, the data show that women comprise less than one quarter of farm holders and on average have smaller farms than men, but provide almost 50 per cent of the family labor force.[35]

Statistics about the role of women in Asia and Africa are difficult to obtain. According to FAOSTAT, only 30 per cent of the rural population in Africa is economically active in agriculture and 40 per cent in Asia – around 45 per cent being women and 55 per cent men.[36] Yet studies carried out or cited by FAO show totally different numbers, indicating that in non-industrialized countries 60 per cent to 80 per cent of the food is produced by women.[37] In Ghana and Madagascar, women make up about 15 per cent of farm holders, but provide 52 per cent of the family labor force and constitute about 48 per cent of paid workers.[38,39] In Cambodia, just 20 per cent of agricultural land holders are women, but they provide 47 per cent of the paid agricultural force and almost 70 per cent of the labor force on family farms.[40] In the Republic of the Congo, women provide 64 per cent of all agricultural labor and are responsible for about 70 per cent of food production.[41] In Turkmenistan and Tajikistan, women are 53 per cent of the active population in agriculture.[42] There is very little data on the evolution of the contribution of women to agriculture, but their share would likely be growing, since migration is resulting in mostly women and girls picking up the workload of those who leave.[43,44,45]

According to FAO, fewer than two per cent of land holders worldwide are women, but figures vary widely.[46] There is broad consensus, however, that even where land is registered as family or joint property between men and women, men still enjoy much wider powers over it. For example, a common situation is that men can make decisions about the land on behalf of themselves and their spouses, but women cannot. Another impediment is that in giving credit, governments and banks require women to present some form of authorization from their husbands or fathers, while men encounter no such barrier. It is no surprise, then, that available data show that only 10 per cent of agricultural loans go to women.[47]

Additionally, inheritance laws and customs often work against women. Males tend to have priority or outright exclusivity in the inheritance of land. In many countries, women can never gain legal control over land, with authority passing to their sons if they are widowed, for example.

The data above support the contention that women are the main food producers on the planet, although their contribution remains ignored, marginalized and discriminated against.

Reversing the trend: give small farmers the means to feed the world

As the data show, land concentration in agriculture is reaching extreme levels. Today, the vast majority of farming families have less than two hectares to feed themselves and humankind. And the amount of land they have access to is shrinking. How are small farmers supposed to sustain themselves in these conditions?

Most families that depend on a small farm need to have family members working outside the farm in order to be able to stay on the land. This situation is often described euphemistically as "diversification", but in reality it means accepting low wages and bad working conditions. For rural families in many countries, it means mass migration leading to permanent insecurity both for those who leave and for those who stay.

If this land concentration process continues, then no matter how hardworking, efficient and productive they are, small farmers will not be able to carry on. The concentration of fertile agricultural land in fewer and fewer hands is directly linked to the increasing number of people going hungry. Genuine land reform is not only necessary, it is urgent. And it must carried out in line with the needs of peasant families and small producer communities. One of these needs is that land be redistributed to small farmers as an inalienable good, not as a commercial asset that can be lost if rural families are not able to cope with the highly discriminatory situations that they face. Farming communities should also be able to decide by and for themselves, and without pressure, the type of land tenure they want to practise.

The situation facing women farmers also requires urgent action. Many international agencies and governments are currently discussing these issues. Land access for women was specifically part of the Millennium Development Goals. The FAO has written numerous documents advocating women's rights over land and agricultural resources. The issue is being considered by the UN Development Programme, the World Bank, the Gates Foundation, the G8 and the G20, among others. However, what these institutions are advocating is often different from what women farmers and women's organizations have been struggling for. These institutions often advocate a system of land rights based on individual property titles that can be bought and sold or used as collateral. This is likely to lead to further concentration of land, just as the allocation of individual land property rights to men has done historically around the world.[48]

Doing nothing to turn this situation around will be disastrous for all of us. Small farmers – the vast majority of farmers, who tend to be the most productive and who produce most of the world's food – are losing the very basis of their livelihoods and existence: their land. If we do nothing, the world will lose its capacity to feed itself. The message, then, is clear. We urgently need to launch, on a scale never seen before, genuine agrarian reform programs that get land back in the hands of small and peasant farmers.

This chapter is based on a GRAIN report which can be accessed online here: https://www.grain.org/e/4929

1 Da Silva, G. (2014) "Opening speech at the Global Forum on Family Farming", Budapest, 5 March http://tinyurl.com/nmkhffc

2 Lowder, S.K. et al. (2014) "What do we really know about the number and distribution of farms and family farms in the world?" Background paper for The State of Food and Agriculture

3 FAO (2014). Figure quoted on pg 8, April, http://tinyurl.com/qh6ql7l

4 See also: FAO (2014) "Family farmers – feeding the world, caring for the earth", http://tinyurl.com/osuelv8

5 McIntyre, B.D. (ed) (2008) "International assessment of agricultural knowledge, science and technology for development: global report", IAASTD, p.8, http://tinyurl.com/mlmuzqy

6 Wenbiao Cai, a professor at the University of Winnipeg, states in several studies that small farms account for most of the farmland in the non-industrialized world. Other examples include allies of small farmer movements like Miguel Altieri, who says that small farms in Latin America "occupy 34.5% of the total cultivated land" (http://tinyurl.com/qxxxf5u), or Greenpeace, which states that "small-scale farmers form the larger part of global agricultural land". (http://tinyurl.com/p233eef)

7 A number of people generously took time to review and comment on earlier drafts of this report or help us with certain problems. Their inputs were very useful and we are grateful to all of them. They include: Maria Aguiar, Valter Israel da Silva, Thomas Kastner, Carlos Marentes, Pat Mooney, Ndabezinhle Nyoni, Jan Douwe van der Ploeg, Mateus Santos, Chris Smaje and Liz Aldin Wiley

8 When we talk about farmers or peasants in this report, we mean food producers including

people who raise livestock, such as herders or pastoralists, fishers, hunters and gatherers

9 The land distribution dataset compiled by GRAIN is available at http://www.grain.org/attachments/3011/download

10 Hazell, P. (2013) "Is small farm led development still a relevant strategy for Africa and Asia?", http://ppafest.nutrition.cornell.edu/authors/hazell.html

11 EUROSTAT, Statistics in focus (2011), "Large farms in Europe", http://epp.eurostat.ec.europa.eu/cache/ITY_OFFPUB/KS-SF-11-018/EN/KS-SF-11-018-EN.PDF

12 Unless otherwise stated, figures on countries of the European Union are based on the Agricultural Structure Survey of 2007, as published data from the 2010 survey did not allow us to do the necessary calculations.

13 ECVC and HOTL (2013) "Land concentration, land grabbing and people's struggles in Europe", 17 April, www.eurovia.org/IMG/pdf/Land_in_Europe.pdf

14 See World Rainforest Movement (2012) "An overview of industrial tree plantations in the global South: conflicts, trends, and resistance struggles", for a discussion on this.

15 Alexandratos, N. and Bruinsma, J. (2012) "World agriculture towards 2030/2050. The 2012 revision", ESA Working Paper 12-03, www.fao.org/docrep/016/ap106e/ap106e.pdf

16 Corley, R.H.V. (2009) "How much palm oil do we need?" Environmental Science & Policy 12: 134–139 http://www.sciencedirect.com/science/article/pii/S1462901108001196

17 Other agencies like the International Land Coalition-led Land Matrix put the figure at 203 million hectares, but over a 10-year period (2000–2010) http://www.landcoalition.org/sites/default/files/publication/1205/ILC%20GSR%20report_ENG.pdf

18 See http://farmlandgrab.org for a range of published reports and day-to-day accounts.

19 See, for example, Nwanze, K.F. (2011) "Small farmers can feed the world"; UNEP, "Small farmers report"; FAO, "Women and rural employment fighting poverty by redefining gender roles" (Policy Brief 5)

20 Instituto Brasileiro de Geografia e Estadistica – IBGE (2006), "Censo Agropecuario 2006", http://tinyurl.com/m376s82

21 Binswanger-Mkhize, H.P. et al (eds) "Agricultural land redistribution. Toward greater consensus", World Bank

22 Russian Federation Federal State Statistics Service (2011) "Russia in Figures 2011"

23 See, for example: Carter, M.R.(1984) "Identification of the inverse relationship between farm size and productivity: an empirical analysis of peasant agricultural production", Oxford Economic Papers, New Series, 36(1): 131–45

24 IFAD (2001) "Assets and the rural poor" in Rural Poverty Report 2001, http://www.ifad.org/poverty/

25 Cornia, G.A. (1985) "Farm size, land yields and the agricultural production function: An analysis for fifteen developing countries", World Development 13(4): 513–34, http://www.sciencedirect.com/science/article/pii/0305750X85900543

26 Anyaegbunam, H.N. et al. (2010) "Analysis of determinants of farm size productivity among small-holder cassava farmers in south east agroecological zone, Nigeria", African Journal of Agricultural Research 52(21): 2882–885, http://www.academicjournals.org/article/article1380871734_Anyaegbunam%20et%20al.pdf

27 Carter, M.R. (1984) op.cit.

28 The nine countries are Austria, Bulgaria, Greece, Italy, Netherlands, Portugal, Romania, Spain and the United Kingdom. See EUROSTAT, Statistics in focus (2011) op.cit.

29 Czech Republic, Estonia, Ireland, Latvia, Lithuania, Slovakia and Sweden. Ibid

30 These figures are obtained by extrapolating the productivity of small farms to 100% of agricultural land. Source: https://www.grain.org/e/4929

31 Van der Ploeg, J.D. (2014), University of Wageningen, personal communication, 25 March

32 For a discussion of food systems and the climate crisis, see: GRAIN (2011) "Food and climate change, the forgotten link", September 28, https://www.grain.org/article/entries/4357-food-and-climate-change-the-forgotten-link

33 FAOSTAT, http://faostat3.fao.org/faostat-gateway/go/to/download/O/OA/E

34 Government of El Salvador, data from agricultural census

35 EU Agricultural Economic Briefs (2012) "Women in EU agriculture and rural areas: hard work, low profile", Brief No. 7, June , http://ec.europa.eu/agriculture/rural-area-economics/briefs/pdf/07_en.pdf

36 FAOSTAT. Search done within "resources" and "population", using annual time series

37 FAO (2009) "Women and rural employment. fighting poverty by redefining gender roles", Economic and Social Perspectives, Policy Brief 5, http://www.fao.org/3/a-ak485e.pdf

38 Ministry of Food and Agriculture of Ghana (2010) "Agriculture in Ghana. Facts and Figures 2010", http://mofa.gov.gh/site/wp-content/uploads/2011/10/AGRICULTURE-IN-GHANA-FF-2010.pdf

39 Ministry of Agriculture, Livestock and Fishery of Madagascar (2007) "Recensement de l'Agriculture. Campagne Agricole 2004-2005", http://harvestchoice.org/publications/madagascar-recensement-de-lagriculture-ra-campagne-agricole-2004-2005-tome-i-generalite

40 FAO and National Institute of Statistics of Cambodia (2010). National Gender Profile of Agricultural Households, 2010. http://www.fao.org/docrep/012/k8498e/k8498e00.pdf

41 IFAD (2009) "Republic of the Congo. Country strategic opportunities programme", http://www.ifad.org/gbdocs/eb/98/e/EB-2009-98-R-20.pdf

42 FAO, Gender Team for Europe and Central Asia (2011) "The crucial role of women in agriculture and rural development", ppt, 3 March, Regional Informal Consultation, Budapest, http://www.fao.org/fileadmin/user_upload/Europe/documents/WPW/gender_files/Gender_REUprez_en.pdf

43 Danziger, N. (2009) "Rural women and migration", International Organization for Migration, http://www.iom.int/jahia/webdav/shared/shared/mainsite/published_docs/brochures_and_info_sheets/Rural-Women-and-Migration-Fact-Sheet-2012.pdf

44 Dodson, B. et al (2008) "Gender, migration and remittances in Southern Africa", The Southern African Migration Project, Migration Policy Series No. 49, http://www.queensu.ca/samp/sampresources/samppublications/policyseries/Acrobat49.pdf

45 Datta, A. and Mishra, S.K. (2011)"Glimpses of women's lives in rural Bihar: impact of male migration", The Indian Journal of Labour Economics 54(3), repub.eur.nl/pub/34865/metis_178663.pdf

46 Doss, C. et al. (2013) "Gender inequalities in ownership and control of land in Africa. Myths versus reality", IFPRI Discussion Paper, http://www.ifpri.org/publication/gender-inequalities-ownership-and-control-land-africa-myths-versus-reality

47 See FAO Infographic at http://www.fao.org/assets/infographics/FAO-Infographic-Gender-ClimateChange-en.pdf

48 See for example the discussion by Celestine Nyamu-Musembi (2006) in "Breathing Life into Dead. Theories about Property Rights: de Soto and Land Relations in Rural Africa", Institute of Development Studies, http://www.ids.ac.uk/publication/breathing-life-into-dead-theories-about-property-rights-de-soto-and-land-relations-in-rural-africa

2.4 Squeezing Africa dry: behind every land grab is a water grab

Food cannot be grown without water. In Africa, one in three people endure water scarcity and climate change will make things worse. Building on Africa's highly sophisticated indigenous water management systems could help resolve this growing crisis, but these very systems are being destroyed by large-scale land grabs amidst claims that Africa's water is abundant, under-utilized and ready to be harnessed for export-oriented agriculture. The current scramble for land in Africa reveals a global struggle for what is increasingly seen as a commodity more precious than gold or oil: water.

The Alwero River in Ethiopia's Gambela region provides both sustenance and identity for the indigenous Anuak people who have fished its waters and farmed its banks and surrounding lands for centuries. Some Anuak are pastoralists, but most are farmers who move to drier areas in the rainy season before returning to the riverbanks. This seasonal cycle helps to nurture and maintain soil fertility. It also helps structure the culture around the repetition of traditional cultivation practices related to rainfall and rising rivers, as each community looks after its own territory and the waters and farmlands within it.

In recent years, Saudi Arabian companies have been acquiring millions of hectares of land overseas to produce food to ship to domestic markets. Saudi Arabia does not lack land for food production. What's missing in the Kingdom is water, and its companies are seeking it in countries like Ethiopia. One new plantation in Gambela, owned by the Saudi-based billionaire Mohammed al-Amoudi, is irrigated with water diverted from the Alwero River. Thousands of people depend on the Alwero's water for their survival and Al-Amoudi's industrial irrigation plans could undermine their access to it. In April 2012, tensions spilled over when an armed group ambushed Al-Amoudi's Saudi Star Development Company operations, leaving five people dead.

The tensions in southwestern Ethiopia illustrate the importance of access to water in the global land rush. Hidden behind the current scramble for land is a worldwide struggle for control over water. Those who have been buying up vast stretches of farmland in recent years, whether they are based in Addis Ababa, Dubai or London, understand that the access to water, often included for free and without restriction, may well be worth more over the longterm than the land deals themselves.

Indian companies, such as Bangalore-based Karuturi Global, are doing the same. Aquifers across the sub-continent have been depleted by decades of unsustainable irrigation. The only way to feed India's growing population, the claim is made, is by sourcing food production overseas, where water is more available.

"The value is not in the land," says Neil Crowder of UK-based Chayton Capital, which has been acquiring farmland in Zambia, "The real value is in water."[1]

And companies like Chayton Capital think that Africa is the best place to find that water. The message repeated at farmland investor conferences around the globe is that water is abundant in Africa and ready to be harnessed. The reality is that one third of Africans already live in water-scarce environments and climate change is likely to increase these numbers significantly. Massive land deals could rob millions of people of their access to water and risk the depletion of the continent's most precious freshwater sources.

All of the land deals in Africa involve large-scale, industrial agriculture operations that will consume massive amounts of water. Nearly all of them are located in major river basins. They occupy fertile and fragile wetlands, or are located in more arid areas that can draw water from major rivers. In some cases the farms directly access ground water by pumping it up. These water resources are lifelines for local farmers, pastoralists and other rural communities. Many already lack sufficient access to water for their livelihoods. If there is anything to be learnt from the past, it is that such mega-irrigation schemes can not only put the livelihoods of millions of rural communities at risk, but also threaten the freshwater sources of entire regions.

When the Nile runs dry....

Africa's longest river, the Nile, is already a source of significant geopolitical tensions aggravated by the numerous large-scale irrigation projects in the region. In 1959, Great Britain brokered a colonial deal that divided the water rights between Egypt and Sudan. Egypt was allocated three-quarters of the average annual flow, while Sudan was allocated one quarter. Massive irrigation schemes were built in both countries to grow cotton for export to the UK. In the 1960s, Egypt built the mighty Aswan dam to regulate the flow of the Nile in Egypt, and increase opportunities for irrigation. The dam achieved those goals, but also stopped the flow of nutrients and minerals that fertilized the soil of Egypt's farmers downstream.

In Sudan, the Gulf States financed a further increase of irrigation infrastructure along the Nile in the 1960s and 1970s in an effort to turn Sudan into the "breadbasket of the Arab world". This was unsuccessful and half of Sudan's irrigation infrastructure currently lies abandoned or underused. Both Sudan and Egypt produce most of their food from irrigated agriculture, but both also face serious problems with soil degradation, salinization, water logging and pollution induced by the irrigation schemes. As a result of all these interventions, the Nile barely delivers water to the Mediterranean any longer – instead, salty seawater now backs into the Nile delta, undermining agricultural production.

The economically, ecologically and politically fragile Nile basin is now the target of a new wave of large-scale agriculture projects. Three of the main countries in the basin – Ethiopia, South Sudan and Sudan – have together already leased out millions of hectares in the basin, and are putting more on offer. To bring this land into production, all of it will need to be irrigated. The first question that should be asked is whether there is enough water to do this. But none of those involved in the land deals, be it the land grabbers or those offering lands to grab, seem to have given the question much thought. The assumption is that there is plenty of water and the newcomers can withdraw as much as they need.

Ethiopia is the source of some 80 per cent of the Nile water. In its Gambela region, on the border with South Sudan, corporations such as Karuturi Global and Saudi Star are already building big irrigation channels that will increase Ethiopia's withdrawal of water from the Nile enormously. These are only two of the actors involved. One calculation suggests that if all the land that the country has leased out is brought under production and irrigation, it will increase the country's use of freshwater resources for agriculture by a factor of nine.[2,3]

Further downstream, in South Sudan and Sudan, some 4.9 million hectares of land has been leased out to foreign corporations since 2006. That is an area greater than the entire Netherlands. To the north, Egypt is also leasing out land and implementing its own new irrigation projects. It remains to be seen how much of all this will actually be brought into production and put under irrigation, but it is difficult to imagine that the Nile can handle this onslaught.

The figures have to be considered with some caution. A limitation of the FAO irrigation figures is that they rely on data provided by individual countries. Criteria on how they were established vary widely – some focus on the available land and others on the available water, yet others on the economic costs. Moreover, the "potential" doesn't take into account that countries upstream might overdraw their water resources, which would affect the amount of water countries downstream would receive. And it remains to be seen whether all the land leased out will actually be brought under production and irrigation as companies pull out, projects collapse or if the land is just being acquired for speculation purposes.

Nevertheless, the FAO figures make it clear that the recent land deals vastly outstrip water availability in the Nile basin. The FAO establishes eight million hectares as the total "maximum value" available for total irrigation in all 10 countries of the Nile basin. But the four countries already have irrigation infrastructure established for 5.4 million hectares and have now leased out a further 8.6 million hectares of land. This would require much more water than is actually available in the entire Nile basin and would amount to nothing less than hydrological suicide.

Water availability is a highly seasonal affair for most people in Africa. But Africa's dry and wet seasons are hidden by the averages and potentials of the quoted figures. Most of the 80 per cent of the Nile water that originates in the Ethiopian highlands falls from the sky and flows into the river between June and August. Local communities have adapted their farming and pastoral systems to make optimum use of the seasonal fluctuations. But the new landowners from abroad want water all year round, with several harvests per year if possible. They will build more canals and dams to make that possible. They also tend to grow crops that need massive amounts of water, such as sugarcane and rice. In all, this means that they'll consume much more than the potentials and averages suggest, putting the FAO figures quoted above in an even more alarming light.

The Niger: another lifeline at risk

Another part of Africa targeted by agribusiness is the land along the Niger River. The Niger is West Africa's largest river, and the third longest in all of Africa, surpassed only by the Nile and the Congo. Millions rely on it for agriculture, fishing, trade and as a primary water source. Mali, Niger and Nigeria are the countries most dependent on the river, but seven other countries in the Niger basin share its water. The river is extremely fragile and has suffered under the strain of dams, irrigation and pollution. Water experts estimate that the volume of the Niger has shrunk by one third during the last three decades alone. Others indicate that the river might lose another third of its flow as a consequence of climate change.[4]

In Mali, the river spreads out into a vast inland delta, which constitutes Mali's main agricultural zone and one of the region's most important wetlands. It is here that the "Office du Niger" is located and where many of the land-grabbing projects are concentrated. The Office du Niger presides over the irrigation of more than 70,000 hectares, mainly for the production of rice. It is the largest irrigation scheme in West Africa, and it uses a substantial part of the river's water, especially during the dry season.

In the 1990s, the FAO estimated Mali's potential to irrigate from the Niger at a bit over half-a-million hectares.[5] But now, due to increased water scarcity, independent experts conclude that the whole of Mali has the water capacity to irrigate only 250,000 hectares.[6] Yet the Malian government has already signed away 470,000 hectares to foreign companies from Libya, China, the UK, Saudi Arabia and other countries in the past few years, virtually all of it in the Niger basin. In 2009, it announced that it would further increase the allowable area of irrigated lands in the country by a mind-boggling one to two million hectares.

A study by Wetlands International calculates that, with the effects of climate change and the planned water infrastructure projects, more than 70 per cent of the floodplains of the Inner Niger Delta will be lost, with a dramatic impact on Mali's ability to feed its people.[7] Those who will suffer the most are the more than one million local farmers and pastoralists in the Inner Niger Delta that now depend on the river and its inner delta for their crops and herds.

Hydro-colonialism?

The Nile and the Niger basins are only two of the examples of the massive giveaway of land and water rights. The areas where land grabbing is concentrated in Africa coincide closely with the continent's largest river and lake systems, and in most of these areas irrigation is a prerequisite of commercial production.

The Ethiopian government is constructing a dam in the Omo River to generate electricity and irrigate a huge sugarcane plantation, a project that threatens hundreds of thousands of indigenous people who depend on the river further downstream. It also threatens to empty the world's biggest desert lake, Lake Turkana, fed by the Omo River. In Mozambique the government signed off on a 30,000-hectare plantation along the Limpopo River which would have directly affected farmers and pastoralists now depending on the water. The project was revoked because the investor didn't deliver, but the government is looking for others to take over. In Kenya, a tremendous controversy has arisen from the government's plans to hand out huge areas of land in the delta of the Tana River, with disastrous implications for the local communities depending on the delta's water. The already degraded Senegal River basin and its delta have been subject to hundreds of thousands of hectares in land deals, putting foreign agribusiness in direct competition for the water with local farmers. The list goes on, and is growing by the day.

Peter Brabeck-Letmathe, the Chair of the Nestlé Group, says that these deals are more about water than land: "With the land comes the right to withdraw the water linked to it, in most countries essentially a freebie that increasingly could be the most valuable part of the deal."[8] Nestlé is a leading marketer of bottled water under brand names including Pure Life, Perrier, S. Pellegrino and a dozen others. It has been charged with illegal and destructive groundwater extraction, and of making billions of dollars in profits on cheap water while dumping environmental and social costs onto communities.[9]

Asked at an agricultural investment conference whether it is possible to make money from water, Judson Hill of one of the private equity funds involved, was unequivocal: "Buckets, buckets of money," he told a meeting of bankers and investors in Geneva. "There are many ways to make a very attractive return in the water sector if you know where to go."[10]

In the not-so-distant future, water will become "the single most important physical-commodity based asset class, dwarfing oil, copper, agricultural commodities and precious metals," says Citigroup's chief economist, Willem Buiter.[11] No surprise, then, that so many corporations are rushing to sign land deals that give them wide-ranging control over African water. Especially when African governments are essentially giving it away. Corporations understand what's at stake. There are "buckets of money" to be made on water, if only it can be controlled and turned into a commodity.

The secrecy that shrouds land deals makes it hard to know exactly what's being handed over to foreign companies. But from those contracts that have been leaked or made public, it is apparent that the contracts tend not to contain any specific mention of water rights at all, leaving the companies free to build dams and irrigation canals at their discretion, sometimes with a vague reference to "respecting water laws and regulations".[12] This is the case in the agreements signed between the Ethiopian government and both Karuturi and Saudi Star in Gambela, for example. In some contracts, a minor user fee is agreed upon for the water, but without any limitation on the amount of water that can be withdrawn. Only in rare cases are even minimal restrictions imposed during the dry season, when access to water is so critical for local communities. But even in instances where governments may have the political will and capacity to negotiate conditions to protect local communities and the environment, this is made increasingly difficult due to existing international trade and investment treaties that give foreign investors strong rights in this respect.[13]

Stop the water grab

If this land and water grab is not put to an end, millions of Africans will lose access to the water sources they rely on for their survival. They may be moved out of areas where land and water deals are made or their access to traditional water sources may simply be blocked by newly built fences, canals and dikes. This is already happening in Ethiopia's Gambela, where the government is forcibly moving thousands of indigenous people out of their traditional territories to make way for export agriculture. As the bulldozers move into the newly acquired lands, this will become an increasingly common feature in Africa's rural areas, generating more tensions and conflicts over scarce water resources.

But the impacts will run far beyond the immediately affected communities. The recent wave of land grabbing is nothing short of an environmental disaster in the making. There is simply not enough water in Africa's rivers and water tables to irrigate all the newly acquired land. If and when they are put under production, these 21st century industrial plantations will rapidly destroy, deplete and pollute water sources across the continent. Such models of agricultural production have generated enormous problems of soil degradation, salinization and waterlogging wherever they have been applied.

India and China, two shining examples that Africa is being pushed to emulate, are now in a water crisis as a result of their Green Revolution practices. More than 200 million people in India and 100 million in China depend on foods produced by the over-pumping of water.[14] Fearing depleted water supplies or perhaps depleted profits, companies from both countries are looking now to Africa for future food production.

Africa is in no shape for such an imposition. More than one in three Africans live with water scarcity, and the continent's food supplies are set to suffer more than any other's from climate change. Building Africa's highly sophisticated and sustainable indigenous water management systems could help resolve this growing crisis, but these are the very systems being destroyed by land grabs.

Advocates of the land deals and mega-irrigation schemes argue that these big investments should be welcomed as an opportunity to combat hunger and poverty in the continent. But bringing in the bulldozers to plant water-intensive export crops is not and cannot be a solution to hunger and poverty. If the goal is to increase food production, then there is ample evidence that this can be most effectively done by building on the traditional water management and soil conservation systems of local communities.[15] Their collective and customary rights over land and water sources should be strengthened not trampled.

But this is not about combating hunger and poverty. This is theft on a grand scale of the very resources – land and water – which the people and communities of Africa must themselves be able to manage and control in order to face the immense challenges of this century.

Box 1: Water mining, the wrong type of farming

If history has anything to teach us, it is that the industrial agriculture being promoted across Africa and the rest of the world is simply not sustainable. In Pakistan, the British Empire built the largest single irrigated area in the world to produce the raw material for the cotton mills in the UK. After independence, irrigation works were extended by the new government, backed by generous funding from the World Bank. Today, 90% of the country's crops - including massive fields of rice and wheat using the

---

technologies of the Green Revolution of the 1960s – are irrigated by the mighty Indus River.

But there was a price to pay. The Indus carries 22 million tonnes of salt each year, but discharges only 11 million tonnes at its exit into the Arabian Sea. The rest, almost a tonne per year for every irrigated hectare, stays on farmers' fields, forming a white crust that kills the crops. One tenth of Pakistan's fields are no longer usable for agriculture, one fifth are badly waterlogged and one quarter produce only meagre crops. Moreover, the water use is so intense that in many years the Indus no longer flows all the way to the sea.

Across the border, in India, deep boreholes watered thirsty new varieties and crops that replaced the indigenous farming systems during the Green Revolution. This raised the country's groundwater consumption to dangerous and totally unsustainable levels. Recent estimates put India's annual abstraction for irrigation at 250 cubic kilometers per year, 40% more than what is replaced by rains. As a result, India's underground water reserves are plunging, forcing farmers to drill deeper every year. In the US, the maize and soybean plantations that dominate the country's midwest have already caused the water table to fall substantially. California, with its vast fruit plantations, pumps 15% more water than the rains replenish.

But perhaps the situation is nowhere more dramatic than in the Middle East. Saudi Arabia has no rain or rivers to speak of, but possesses vast "fossil water" aquifers beneath the desert. During the 1980s the Saudi government invested $40 billion of its oil revenues to pump this precious water to irrigate a million hectares of wheat. Later, in the 1990s, in order to feed the growing industrial dairy farms that popped up across the desert, many farmers switched to alfalfa, a crop that needs even more water. It was clear that the miracle couldn't last; the aquifers soon collapsed and the government decided to outsource its food production to Africa and other parts of the world instead. Some 60% of the country's fossil water under the desert was squandered in the process. Gone and lost forever.

Data derived from Fred Pearce's excellent book on the global water crisis
When the Rivers Run Dry, Eden Project Books, 2006.

Table 1 Selected African land deals and their water implications

Land deal summary	Water implications
Mozambique – Limpopo River	
30,000 hectares close to Massingir Dam leased to Procana for sugarcane production. Project was suspended and government is now looking for new investors. One study puts the total new irrigation plans due to the various land acquisitions at 73,000 hectares.	One study concluded that the Limpopo River does not carry sufficient water for all planned irrigation and that only about 44,000 hectares of new irrigation can be developed, which is 60% of the envisaged developments. Any additional water use would certainly impact downstream users and thus create tensions.[16]
Tanzania – Wami River	
Eco-Energy has been granted a concession of 20,000 hectares to grow sugarcane. The company claims that the size of the project has now been reduced to 8,000 hectares.	The Environmental Impact Assessment (EIA) for the project revealed that the amount of water Eco-Energy requested to withdraw from Wami River for irrigation during the dry season was excessive and would reduce the flow of the river. The EIA also predicts an increase in local conflicts related to both water and land.[17]
Kenya – Yala Swamp (Lake Victoria)	
Dominion Farms (US) established its first farm on a 7,000–hectare piece of land in the Yala Swamp area in Kenya, which it obtained on a 25-year lease.	The local communities living in the area complain of being displaced without compensation, of losing access to water and pasture for their livestock, of losing access to potable water and of pollution from the regular aerial spraying of fertilizers and agrochemicals. They continue to struggle to get their lands back and to get Dominion to leave.[18]

Land deal summary	Water implications

Ethiopia/Kenya – Omo River and Lake Turkana

The Ethiopian government is building an enormous dam in the Omo River to produce electricity and to irrigate 350,000 hectares for commercial agriculture, including 245,000 hectares for a huge state-run sugarcane plantation. Known as 'Gibe III', the dam has sparked tremendous international opposition due to the environmental damage it will cause, and the impact it will have on indigenous people depending on the river.

Descending from the central Ethiopian plateau, the Omo River meanders across Ethiopia's southwest before spilling into Kenya's Lake Turkana, the world's largest desert lake. The Omo River and Lake Turkana are a lifeline for over half a million indigenous farmers, herders and fishers, and the Gibe III Dam now threatens their livelihood. Construction of the dam began in 2006. Studies suggest that irrigating 150,000 hectares would lower Lake Turkana by eight meters by 2024. If 300,000 hectares are irrigated, the lake level will decline by 17 meters, threatening the very future of the lake, which has an average depth of only 30 meters.[19]

Ethiopia – Nile River[20]

Multiple foreign investors, including the following in the Gambela region: Karuturi Global Ltd from India which got a 50-year renewable lease on 100,000 hectares with an option for another 200,000 hectares; Saudi Star leased 140,000 hectares and is trying to get more.
Ruchi Group from India signed a contract for a 25-year lease on 25,000 hectares in the same area.

Ethiopia has leased out some 3.6 million hectares. The vast majority of these are in the Nile basin, including the Gambela region. The FAO puts the irrigation potential of the Nile basin in Ethiopia at 1.3 million hectares. So if all the land offered for lease is brought into production and under irrigation, the plantations will draw more water than the Nile can handle. The first ones to lose out are the local communities. The government has started a villagization program in which it is forcibly relocating approximately 70,000 indigenous people from the western Gambela region to new villages that lack adequate food, farmland, healthcare, and educational facilities.

Land deal summary	Water implications
Sudan and South Sudan – Nile River	
Multiple investors, including Citadel Capital (Egypt), Pinosso Group (Brazil), ZTE (China), Hassad Food (Qatar), Foras (Saudi Arabia), Pharos (UAE), and others. Total land deals documented by GRAIN amount to 3.5 million hectares in Sudan, and 1.4 million hectares in South Sudan.	Together Sudan and South Sudan have some 1.8 million hectares under irrigation, virtually all of it drawing from the Nile. FAO calculates that, together, Sudan and South Sudan have an irrigation potential of 2.8 million hectares. But GRAIN identified almost 4.9 million hectares that have been leased out to foreign investors in these two countries since 2006. Of course, considering the recent tense political situation, it remains to be seen whether and when this land will be put under production. But even if a part of it is, there is clearly not enough water in the Nile to irrigate it all.
Egypt – Nile River	
GRAIN documented the acquisition of some 140,000 hectares of farmland by Saudi and UAE agribusiness in Egypt for food and fodder for export by Al Rajhi and Jenat (Saudi Arabia), Al Dahra (UAE) and others.	Egypt is fully dependent on the water of the Nile for its food production. Currently the country has some 3.4 million hectares under irrigation, and FAO calculates that it has an irrigation potential for 4.4 million hectares. It still has to import much of its food. The country is continuously expanding its agricultural area, including the Toshka project to transform 234,000 hectares of Sahara desert into agricultural land in the South, and the Al Salam Canal to irrigate 170,000 hectares in the Sinai, Despite concerns over the needs for water to feed its own population, the Egyptian government has signed off to lease at least 140,000 hectares to agribusiness from the Gulf States to produce food and feed for export. It is difficult to see how this is compatible with feeding its own population.

Land deal summary	Water implications
Kenya – Tana River Delta	
The government has given tenure rights and ownership of 40,000 hectares of Tana Delta land to TARDA (Tana River Development Authority) which entered into a joint venture with Mumias Sugar company to establish sugarcane plantations. A second sugar company, Mat International, is in the process of acquiring more than 30,000 hectares of land in Tana Delta and another 90,000 hectares in adjacent districts. The company has not carried out any environmental or social impact assessments. Bedford Biofuels Inc, from Canada, is seeking a 45-year lease agreement on 65,000 hectares of land in Tana River District to transform it into biofuel farms, mainly growing Jatropha.	The Tana is Kenya's largest river. Its delta covers an area of 130,000 hectares and is one of Africa's most valuable wetlands. It is home to two dominant tribes, the Orma pastoralists and the Pokomo agriculturalists. According to one study, more than 25,000 people living in 30 villages stand to be evicted from their ancestral land that has now been given to TARDA. The impacts of these intensive agricultural projects are numerous and they raise both environmental and social issues. Even the Environmental Impact Assessment of Mumias questions whether the proposed abstraction of irrigation water from the Tana River can be maintained during dry months and drought periods. Reduced flow could lead to damage of downstream ecosystems, reduced availability for livestock and wildlife and increased conflict, both inter-tribal and between humans and wildlife.[21]
Mali – Inner Niger Delta[22]	
GRAIN has documented the acquisition of some 470,000 hectares of farmland in Mali by different corporations from all over the world. They include Foras (S. Arabia); Malibya (Libya); Lonrho (UK); MCC (US); Farmlands of Guinea (UK), CLETC (China) and several others. Virtually all of this is in the "Office du Niger", located in the Inner Niger Delta, a huge inland delta which constitutes Mali's main agricultural area.	The FAO puts Mali's potential to irrigate from the Niger at about half a million hectares. But due to increased water scarcity, independent experts conclude that Mali has the water capacity to irrigate only 250,000 hectares. The government has already signed away rights to 470,000 hectares in the delta – all of it to be irrigated. And it announced that up to 2 million hectares more are available. One study by Wetlands International calculates that the combined effects of climate change and all the planned water infrastructure projects will result in the loss of more than 70% of the floodplains of the delta.

Land deal summary	Water implications
Senegal – Senegal River basin	
GRAIN has documented the acquisition of some 375,000 hectares of farmland by investors from China (Datong Trading), Nigeria (Dangete Industries), S. Arabia (Foras), France (SCL) and India.	A lot of the land deals are in the basin of the Senegal River, which is the main irrigated rice-producing area of Senegal. Around 120,000 hectares in the area are suitable for irrigated rice production and about half of these are currently being farmed under irrigation. The FAO calculates that the river has a total irrigation potential of 240,000 hectares. UNESCO reports that the flood plain ecosystems of the Senegal River are in bad shape due to dam building: "In less than 10 years, the degradation of these environments and the consequences on the health of the local population have been dramatic." Taking more water from the river to produce export crops will make a bad situation worse.[23]
Cameroon	
The agro-industrial group Herakles American Farms leased more than 73,000 hectares of farmland in southwest Cameroon to produce oil palm.	According to local NGOs, the contract gives the company "the right to use, free, unlimited quantities of water in its land grant". It concludes that from a contractual standpoint the company clearly has priority over local communities when accessing water, and fears that the environmental and socio-economic impact will be severe. In 2011, the local youth took to the streets to block the bulldozers in protest. The Mayor of Toko, which is in the area affected by the land deal, drew attention to its impact on the country's major watershed: "This particular area is one of the most important watersheds of Cameroon. We don't need SG SOC or Herackles farms in our area."[24]

Full article & references: https://www.grain.org/e/4516

1 Neil Crowder, CEO Chayton Africa, Zamiba Investment Forum (2011), http://vimeo.com/38060966
2 Oakland Institute (2011) "Landgrabs leave Africa thirsty", December, http://www.oaklandinstitute.org/land-deal-brief-land-grabs-leave-africa-thirsty
3 Estimates of land deal figures for Ethiopia vary wildly. Here we use 3.6 million hectares, as indicated in the 2011 Oakland Institute's country report on the issue. See http://tinyurl.com/br8jz7s. In 2012, Ethiopia's Prime Minister Meles Zenawi announced that the country had made available 4 million hectares to agricultural investors. See http://farmlandgrab.org/post/view/20468
4 Pearce, F. (2006) *When the Rivers Run Dry*, Eden Project, p. 146.
5 FAO (1997) "Irrigation potential in Africa: A basin approach", http://www.fao.org/docrep/W4347E/w4347e0i.htm#the niger river basin
6 Quoted in SIWI (2012) "Land acquisitions: How will they impact transboundary waters?", http://www.siwi.org/sa/node.asp?node=1440
7 Zwarts, L. (2010) "Will the Inner Niger Delta shrivel up due to climate change and water use upstream?", Wetlands International, http://www.wetlands.org/WatchRead/tabid/56/mod/1570/articleType/ArticleView/articleId/2687/Default.aspx
8 Foreign Policy (2009), "The next big thing: H_2O", 15 April, http://www.foreignpolicy.com/articles/2009/04/15/the_next_big_thing_h2o
9 In 2001, residents of the Serra da Mantiqueira region of Brazil, investigating changes in the taste of their water and the complete dry-out of one of their springs discovered that Nestlé/Perrier was pumping huge amounts of water from a 150m well in a local Circuito das Aguas, or "water circuits" park whose groundwater has a high mineral content and medicinal properties. The water was being demineralized and transformed into table water for Nestlé's Pure Life brand. Water usually needs hundreds of years inside the earth to be slowly enriched by minerals. Overpumping decreases its mineral content for years to come. Demineralization is illegal in Brazil, and after the Movimento Cidadania pelas Águas mobilized, a federal investigation was opened resulting in charges against Nestlé/Perrier. Nestlé lost the legal action, but continued pumping water while it fought the charges through appeals. See http://www.corporatewatch.org.uk/?lid=240#water
10 MacInnes, L. (2010) "Private equity sees 'buckets of money' in water buys", *Reuters*, 9 November, http://www.reuters.com/article/2010/11/09/us-farming-investing-water-idUSTRE6A82ZV20101109
11 Quoted in *Financial Times* (2011) "Willem Buiter thinks water will be bigger than oil", 21 July, http://ftalphaville.ft.com/blog/2011/07/21/629881/willem-buiter-thinks-water-will-be-bigger-than-oil/
12 For access to the contracts that we have been able to get hold of, see: http://farmlandgrab.org/home/post_special?filter=contracts
13 The issue of land and water rights in the context of international trade and investment treaties is further discussed in: Smaller, C. and Mann, H. (2009) "A thirst for distant lands", IISD, http://www.fao.org/fileadmin/templates/est/INTERNATIONAL-TRADE/FDIs/A_Thirst_for_distant_lands.pdf
14 Pearce, F. (2006) *op.cit.* See also Water Mining in GRAIN (2014) "The many faces of land grabbing", https://www.grain.org/article/entries/4516-squeezing-africa-dry-behind-every-land-grab-is-a-water-grab?print=true#box 1
15 For more details and examples, see: Oakland Institute (2011) op.cit.
16 Van der Zaag, P. et al. (2010) "Does the Limpopo River Basin have sufficient water for

massive irrigation development in the plains of Mozambique?", http://www.sciencedirect.com/science/article/pii/S1474706510001555

17 Oakland Institute (2011) *op.cit.*

18 GRAIN (2012) Dataset on landgrabbing at http://www.grain.org/e/4479

19 See: International Rivers on the Gibe III Dam website: http://www.internationalrivers.org/africa/gibe-3-dam-ethiopia. Also Oakland Institute (2011) *op.cit.*

20 For sources on the countries in the Nile basin see the original article on which this article is based at https://www.grain.org/e/4516

21 Sources include: tanariverdelta.org, http://www.tanariverdelta.org/tana/g1/projects.html; Tember, L. (2009) "Let them eat sugar: life and livelihood in Kenya's Tana Delta", http://tinyurl.com/cdlcspn; and Nunow, A.A. (2011) "The dynamics of land deals in the Tana Delta, Kenya", http://tinyurl.com/d42rfqf

22 For sources on the Niger basin, see the original article on which this article is based at https://www.grain.org/e/4516

23 GRAIN (2012) *op.cit.*; FAO, Aquastat *op.cit.*; and Unesco "Senegal River Basin", http://webworld.unesco.org/water/wwap/case_studies/senegal_river/

24 Infosud (2012) "Cameroun: les terres de la discorde louées aux Américains", http://tinyurl.com/c82ae2m and Nganda Valentine Beyoko, Mayor of Toko Council, personal communication, 26 March 2012

2.5 Asia's agrarian reform in reverse: laws taking land out of small farmers' hands

Asia is a land of small farmers. But across the continent, governments are introducing changes to land laws that threaten to displace millions of peasants and undermine local food systems. The region is witnessing agrarian reform in reverse.

Despite decades of rapid economic growth and industrialization, there are still more small farmers in the Asian countryside than in the rest of the world combined. But small farmers in Asia are being squeezed onto ever smaller parcels of land. Across the continent, farmland is being gobbled up for dams, mines, tourism projects and large-scale agriculture, with scant regard for the people living off those lands. Farms that peasant families have cared for for generations are being paved over for new highways or real estate development as cities expand. Long-standing government promises to redistribute land more fairly have been broken – in many places, governments are taking land away from peasant farmers.

Land concentration in Asia is higher now than it has ever been. Just six per cent of Asia's farm owners hold around two-thirds of its farmland. Many of these landowners are politically connected elites, as is the case in the Philippines, Cambodia, Malaysia, Pakistan, and Indonesia.[1]

As this concentration increases, one consequence is the eruption of conflicts over land throughout the continent. Peasant protests against land grabs have become a regular sight on the streets of major cities like Phnom Penh and Manila. The court systems in China and Vietnam are backlogged with thousands of rural land conflict cases. And militarized repression is a harsh daily reality in many places where communities are resisting land grabbing, from West Papua to West Bengal.

Governments across Asia are quietly proceeding with a raft of legislative changes to remove the few protections that small farmers have traditionally enjoyed, exposing them to the takeover of their lands for large-scale corporate farming. The changes differ from country to country, but they are all designed to make it easier for companies to acquire large areas of land from small farmers.

These legislative changes will displace millions of peasant families, undermine local food systems and increase violent conflicts over land.

Forcing small farmers off their lands through legal reform

The legislative push to transfer land from small farmers to corporations is prompted by growing interest in farmland. Big money is flowing into plantation companies and other corporate farming ventures from banks, hedge funds, tycoons and transnational commodity traders. Governments are under pressure from these investors to open up agricultural land, and few are putting up any resistance. The spree of bilateral and multilateral free trade agreements that Asian governments have signed on to over the past decade and a half has locked countries into policies that favor corporate farming and foreign investors over small-scale producers (*See Box 1: Trade agreements and land transfers on page 98*).

In some cases the legislation is geared mainly towards a transfer of lands for industrial, tourism or infrastructure purposes, not corporate farming, but the clear trend across the region is the removal of legislative and other impediments that prevent foreign and national companies from acquiring large areas of farmland.

Each country has a different approach, but the changes can be loosely grouped into two types. On the one hand, there are laws or policies that enable governments to carve up large areas of land into concessions and lease or sell these to the companies. This is the case in Burma, Cambodia, Laos, Indonesia, Pakistan, Papua New Guinea and Thailand. On the other hand, laws are being passed or amended to legalize new schemes that consolidate small farms and transfer the lands to companies engaged in corporate farming. Each scheme comes with a different label, such as agri-parks in India, land circulation trusts in China, banks in Korea, clusters in the Philippines or special agricultural production companies in Japan. By GRAIN's calculations, the legislative changes have already led to the transfer from small farmers to agribusiness companies of at least 43.5 million hectares of farmland in Asia.[2]

The number of small farmers in Asia is shrinking, as is the size of their landholdings, while the number of corporate farms is growing rapidly. For example, the number of small farmers in Indonesia fell by 16 per cent between 2003 and 2013, while the number of large-scale farms increased by 54 per cent and the number of plantations increased by 19 per cent over the same period. Most of Indonesia's farmers, around 55 per cent, now farm on less than half a hectare.[3] Similarly, the number of Japanese farmers has dropped by 40 per cent since 2000, while the number of "agricultural production companies" has increased to 14,333, double what it was in 2004.[4]

Box 1: Trade agreements and land transfers

Free trade and investment agreements play an important role in bringing about laws and policies that facilitate the transfer of lands from small farmers to big agribusiness. They do so both indirectly, by encouraging specialized, vertically integrated production of export commodities, and directly, by obliging governments to remove barriers to foreign investment, including in agriculture.[5]

For example, Australia's free trade negotiations with China triggered a rapid and massive flow of investment from Chinese companies into Australian farmland for the production of export commodities such as dairy, sugar and beef. The investment was so controversial that the government was compelled to negotiate a new regulation requiring its Foreign Investment Review Board to scrutinize farmland sales to foreign buyers that exceed a cumulative $15 million. That condition was written into the FTAs negotiated with China, Korea and Japan, in 2014, but it does not apply to foreign companies from the US, New Zealand or Chile because FTAs with these countries had already been concluded.[6]

In Cambodia, the adoption of the Economic Land Concession (ELC) law in August 2001 is intimately connected to the "Everything But Arms" (EBA) preferential trade scheme that it signed with the European Union a few months earlier in March 2001. The ELC established a legal framework for granting large-scale, long-term land concessions of up to 10,000 hectares for up to 99 years for the development of industrial agriculture. Several plantation concessions have since been awarded to companies for the production of sugar exports to Europe under the EBA.

A 2013 assessment found that Cambodia's large-scale land concessions policy and the EU's EBA were together responsible for devastating human rights impacts. In Koh Kong province, for example, two villages and more than 11,500 hectares of rice fields and orchards belonging to more than 2,000 families were destroyed to make way for a sugarcane plantation. More than 1,000 men, women and children were left homeless.[7]

In Japan, the government's decision to revise its agriculture land law was tightly connected to its participation in the Trans Pacific Partnership (TPP), which was concluded in February 2016. In the lead-up to it, Japan was already preparing for an eventual decrease in tariffs on agricultural produce by encouraging a shift from small-scale farming to corporate farming, which it views as more internationally competitive. The government is thus loosening various regulations on the entry of private-sector firms into farming, encouraging farmland consolidation and establishing two strategic special zones for corporate farming.[8] These zones will include approximately 1.5 million hectares of farm land - one third of the total 4.6 million hectares currently under cultivation in the country.[9]

A massive transfer of lands

The arguments used to justify modifying or changing land laws do not hold up to scrutiny. Peasants are said to be abandoning the countryside in favor of work in the cities. We are told that large farms are more efficient and competitive and that corporate farming creates jobs.[10] Liberalized land markets, say donors and international lenders, create social stability and stimulate economic development.[11]

Growing adoption of industrial farming systems and increasing corporate control of distribution of food – changes supported by the new land laws – have led to a reliance on expensive inputs, the degradation of land and biodiversity and volatile price changes for produce. The impact on peasant farmers has been catastrophic, in some places triggering a wave of suicides among indebted farmers forced to give up their land.

Farmers across Asia are fighting for their land, not trying to flee it. If anything, it is the policies of Asia's governments that are creating conditions that compel peasants to migrate to urban areas to provide a cheap labor supply for export manufacturing.

The arguments about productivity and efficiency are also false. Asia's small farmers are among the most efficient and productive farmers in the world. The truth is that small farmers feed Asia. Despite having the highest percentage of small farms, Asian farmers are able to produce 44 per cent of total world production of cereal. India is the largest dairy producer in the world with 85 per cent of the national dairy sector handled by small-scale and backyard dairy farmers. China's backyard farmers, rearing between one and 10 pigs per year, account for 27 per cent of nationwide pork production. And five Asian countries with a majority of small farmers, China, India, Indonesia, Thailand and Vietnam, account for 70 per cent of global rice production.[12]

Resistance

The transfer of land in Asia represents a fundamental shift away from small-scale, traditional agriculture and local food systems to a corporate food chain supplied by industrial agriculture. If they are allowed to go forward, these changes will have major impacts on everything from food safety to the environment, from local cultures to people's livelihoods.

Governments are choosing sides in a struggle over the future of land and food. The case of India is a good example: sustained popular mobilization led the government to adopt legislation that required social impact legislation and broad consultation with affected communities before transferring land, but the land acquisition act of 2013 had barely been passed before a land ordinance that overturned it was rushed through by the executive at the end of 2014.

The land struggles that rural people are engaged in are taking on much larger social dimensions. This can be seen in the street protests against the land acquisition ordinance in India, or the creative actions to stop the conversion of farmlands in Taiwan. People across Asia are making it clear that they want farmland to remain with their farmers. They want their governments to stop facilitating a corporate take-over of agriculture.

The example of resistance in India – and powerful popular mobilizations around this issue are also taking place in Cambodia, Taiwan, the Philippines and elsewhere – shows the strength of joint efforts by rural and urban communities, as well as the importance of links between the local and regional levels in building effective political pressure.

There is an urgent need to further strengthen this resistance to the co-option of land reform in the interests of agro-industry. Farmers, indigenous groups and civil society organizations across the region are building coalitions to defend peasants' interests against trade agreements and national policies that facilitate the privatization and commodification of farmland.

Hungry for land

(GRAIN would like to acknowledge and thank everyone who contributed their thoughts, knowledge and experience to this chapter: Yan Hairong, Forest Zhang, Assembly of the Poor – Thailand, Equitable Cambodia, the India Coordination Committee of Farmers Movement, NOUMINREN, Taiwan Rural Front, Peoples Common Struggle Centre – Pakistan, Lao land issue working group)

Full report and references: https://www.grain.org/e/5195

1 GRAIN (2014) "Hungry for Land", http://www.grain.org/article/entries/4929-hungry-for-land-small-farmers-feed-the-world-with-less-than-a-quarter-of-all-farmland

2 GRAIN calculated the numbers from existing government projects and publicly announced plans, but with limited data available in several countries, the total amount of land that has been transferred is likely much higher.

3 Sensus Pertanian Indonesia (2013), www.st2013.bps.go.id/dev/st2013/

4 Shimbun, Y. (2014) "Rebuilding agriculture in Japan", 10 December, http://the-japan-news.com/news/article/0001775418

5 Cotula, L. (2013) "Tackling the trade law dimension of land grabbing", International Institute for Environment and Development, 14 November, http://www.iied.org/tackling-trade-law-dimension-land-grabbing

6 Dakis, S. (2015) "Trade Minister defends tighter foreign investment scrutiny", ABC Rural, 15 February, http://farmlandgrab.org/post/view/24537-trade-minister-defends-tighter-foreign-investment-scrutiny

7 Equitable Cambodia and Inclusive Development (2013) "Bittersweet harvest: a human rights impact assessment of the European Union's Everything But Arms initiative in Cambodia", http://www.inclusivedevelopment.net/wp-content/uploads/2013/10/Bittersweet_Harvest_web-version.pdf

8 Shimizu, K. and Mclachlan, P. (2014) "Showdown: the Trans-Pacific Partnership vs. Japan's farm lobby", *The National Interest*, 2 October, http://nationalinterest.org/blog/the-buzz/showdown-the-trans-pacific-partnership-vs-japans-farm-lobby-11394

9 31st policy recommendation of the Policy Council, Japan Forum on International Relations (2009) "Japan's strategy for its agriculture in the globalised world", www.jfir.or.jp/e/pr/pdf/31.pdf

10 See, for instance the FAO's work on land tenure and administration in the Asia-Pacific region in co-operation with UN Habitat, the World Bank, IFAD, and UN Economic Commission for Europe (UNECE) in 2008 and at the FAO's 32nd regional conference for Asia and Pacific roadmap in 2014, www.fao.org/about/meetings/aprc32/en/. In Burma, Lao PDR, and China, schemes are promoted for reducing rural poverty where small farmers are promised shares in the company's annual profit or waged labor as a "reward" for giving up their lands.

11 Grimsditch, M. and Henderson, N. (2009) "Untitled: Tenure insecurity and inequality in Cambodia land sector", Bridges Across Borders Southeast Asia, Centre on Housing Rights and Eviction and Jesuit Refugee Services, October, www.babcambodia.org/untitled/

12 FAOSTAT (2015), www.faostat3.fao.org

2.6 The land grabbers of the Nacala Corridor

From liberation to land grabs

Mozambique declared independence on 25 June 1975 after a decade of armed struggle. The peasants, workers, and students of Mozambique had defeated the Portuguese empire, guided by a common ideal of "freedom of man and earth". The ideals of the national liberation struggle are enshrined in the republic's first constitution, which recognizes the right of the Mozambican people to resist all forms of oppression.

Land was particularly important to the country's liberation struggle. Portuguese settlers had occupied vast tracts of the country's most fertile lands. When Mozambique achieved independence, these lands were immediately taken back and nationalized. Under the 1975 constitution, the state – on behalf of the Mozambican people – became the owner of all land in the country. The constitution also recognized agriculture as the foundation of development, with industry as its main engine, to be underpinned by a policy of national industrialization led by state companies and co-operatives.

One year after independence, a brutal civil war broke out which ended only with the founding of a second republic in 1992. Then followed two decades of structural adjustment policies imposed by the World Bank and International Monetary Fund (IMF). Today, 40 years after independence, the revolutionary vision of the national liberation movement is in tatters and the Mozambican government is thoroughly dominated by a neoliberal ideology that relies narrowly on foreign investment for the development of all economic sectors, whether agriculture, infrastructure, fishing, tourism, resource extraction, health or education.

Foreign investment in the country has thus expanded rapidly in recent years. According to the National Bank of Mozambique, the net inflow of foreign direct investment (FDI) in 2013 amounted to $5.9 billion, up 15.8 per cent from 2012, making Mozambique the third-largest destination for FDI in Africa.[1] Much of this capital has gone into resource extraction, such as mining and exploration of hydrocarbons. But agriculture is also emerging as an important target of foreign companies, especially in the Nacala Corridor, a vast stretch of fertile lands across northern Mozambique where millions of peasant families live and farm.

Over and above this, these investments are the result of a very strong alliance between international capital through the big transnational corporations, with the support of the governments in their home countries, and the local political-economic elite with the intention of exploiting the country's main agro-ecological regions and the potential in mining and hydrocarbons. This research analyzes the roles of the different players in the occupation and appropriation of the Nacala Corridor, one of the country's richest regions, which, besides being home to its main ecosystems, is the repository of reserves of a number of minerals.

A new era of plantations in northern Mozambique

The rising foreign interest in farmland is not unique to Mozambique. The entire African continent has been hit by a scramble for farmland. Since 2008, foreign companies have been scouring Africa in search of fertile lands to produce agricultural commodities for export. Hundreds of deals have already been signed covering millions of hectares.

The government of Mozambique has unabashedly sought to attract this wave of foreign agricultural investment to its shores, and particularly to the Nacala Corridor in the north of the country. It is partnering with foreign governments and donors, most notably Japan and Brazil, on a massive program known as ProSavana, which aims to transform 14 million hectares of lands currently cultivated by peasant farmers serving local markets in this area into massive farming operations run by foreign companies to produce cheap agricultural commodities for export.

Mozambique's National Peasants Union (UNAC) has been leading a campaign to raise awareness about the situation in the Nacala Corridor and to oppose ProSavana. Strong national and international opposition has helped to slow down the project and derail some of its more aggressive land grabbing components. However, the government and foreign companies have not given up on taking control of the lands and water resources of the Nacala Corridor for large-scale agribusiness.

In January 2014, high-level government officials and entrepreneurs gathered for the presentation of a new development project in the Lúrio River Basin. The $4.2 billion initiative involves a massive farm project along the Lúrio River, at the intersection of the provinces of Niassa, Nampula and Cabo Delgado. The project is being overseen by a company called Companhia de

Desenvolvimento do Vale do Rio Lúrio, which appears to be run by TurConsult Ltda. TurConsult is owned by Rui Monteiro, an influential entrepreneur in Mozambique's hotel and tourism industry, and Agricane, a South African company that has provided consulting and management services to many large-scale agribusiness projects in Africa, especially in the sugar industry. It is not clear who is financing the project. The Companhia de Desenvolvimento do Vale do Rio Lúrio plans to construct two hydroelectric dams on the Lúrio River and to create an irrigation scheme covering 160,000 hectares. Another 140,000 hectares will be developed for rain-fed agriculture, contract farming and livestock production. The project will focus on the export production of cotton, maize, cereals, and cattle, as well as sugarcane for biofuel ethanol. As with ProSavana, the details of this project are being kept hidden from the public, but preliminary estimates are that upwards of 500,000 people living in the area will be affected by the project.

The Lúrio River project and ProSavana should not be seen separately. They are part of a broader push, involving the World Bank and the G8's New Alliance for Food Security and Nutrition, to open Mozambique up to large-scale agribusiness projects.

The G8's New Alliance was proposed by the US government and signed by some 40 states, international financial institutions and multilateral organizations at the 2009 G8 Summit in L'Aquila, Italy. Under the New Alliance, Mozambique has adopted a National Agricultural Investment Plan (PNISA) which has been shaped primarily to serve the interests of the G8 countries and their respective corporations, under the guise of enhancing Mozambique's "food and nutrition security".[2]

The Mozambican government has already instituted significant reforms to facilitate foreign investment in agribusiness. These include changes to land laws to provide a more flexible allocation of land titles, known as a "right of use and benefit of land" (DUAT), and changes to its seed and fertilizer laws to harmonize them with the Southern African Development Community (SADC). These reforms are important in opening the door to mega agribusiness projects in the Nacala Corridor.[3]

Another important project encouraging the scramble for lands in the area is the strategic plan for the Nacala Corridor. This plan pulls together various major investments in infrastructure, resource extraction, mining and transportation. The strategic plan is funded by the Japan International Cooperation

Agency (JICA) – the Japanese company Mitsui is a major investor in the Moatiza coal mine, the railway and the port of Nacala, as well as being a potential investor in agricultural production in the area.[4]

The governments, companies and agencies promoting ProSavana and the other projects in the Nacala Corridor maintain that local farmers will benefit from the new investment, infrastructure and access to markets. They also say that peasants will not be displaced from their lands to make way for corporate farms. Yet it is apparent that these projects are already encouraging land grabs in the Nacala Corridor. A number of foreign companies, some in collaboration with local businesses linked to members of Mozambique's ruling FRELIMO party, have already acquired large tracts of farmland in the area and have displaced thousands of peasant families.

The money that is now pouring into agribusiness in the Nacala Corridor is essentially recreating what the local people experienced under Portuguese colonialism. During the colonial period, the administration generously handed out the most fertile lands in the area to Portuguese investors. At times the Mozambicans farming the lands were given small amounts in compensation, but most often they were simply evicted. With independence in 1975, the Portuguese investors fled and the local people returned to their lands to resume farming. In some cases, state companies took over the colonial plantations, but few of these companies were able to maintain production, and communities later reclaimed much of this land as well.

Mozambique's land law gives communities possession over lands that they have farmed for more than 10 years. So these former colonial estates should now have formally reverted to local farmers. But as the area has once again become a target for foreign investment in agriculture, the Mozambican government is colluding with foreign investors to provide them with long-term leases over these same lands. The colonial echo is strengthened by the fact that some of the investors are Portuguese families that became rich during the colonial period and are now coming back to Mozambique to set up plantations on the very same lands Portuguese colonialists fled 40 years ago. Few of them have backgrounds in agriculture, but many have connections with influential members of the ruling FRELIMO party, who help them acquire lands and manage any opposition from local communities.

Often the communities are not even aware who is grabbing their lands. The companies that take possession are typically registered in offshore tax havens like Mauritius, where the identity of the owners of the companies and the financial records are kept secret. This leaves the Mozambican authorities and affected communities few options to hold these companies to account for their actions or ensure that a minimum amount of their profits stays within the country.

These land grabs provide a clear picture of the kind of "investment" Mozambican peasants can expect from ProSavana, the Vale do Rio Lúrio project and other initiatives to encourage foreign investment in agribusiness in the country.

Box 1: Profile – Mozaco and the Grupo Espirito Santo

The Mozambique Agricultural Corporation (Mozaco) was established in Mozambique in June 2013 by Rioforte Investments and João Ferreira dos Santos (JFS Holding).

Mozaco says it acquired a DUAT for 2,389 hectares near the village of Natuto in the Malema District of Nampula Province in June 2013, where it plans to cultivate soybeans and cotton. The company says its "objective is to expand it up to 20,000 hectares". It also intends to pursue contract production with 116-170 local farmers on 83 hectares, building on a program developed with the US NGO Technoserve.[5]

The area occupied by Mozaco in Natuto community, in Malema District, is an area that in colonial times was occupied by a settler called Morgado, who produced tobacco and cotton on about 1,000 hectares. After independence, the government nationalized the lands and installed a state company known as Unidade de Namele, which also operated farms in Ribaué and Laulaua Districts. At its height, the state farm employed 5,000 workers but, by 1989, with the civil war intensifying, it was shut down.

"When the company was closed, workers were owed several years of back wages," says a 48-year-old father of seven from Natuto who worked at the Unidade de Namele farm. "But, as it was impossible to complain because of the level of government repression at the time, many of us just ended up taking small parcels of land from the state farm of between one to five hectares, which we cultivate to this day. The company João

---

Fereira dos Santos cultivated a few hectares of Virginia tobacco in the early 1990s, but it abandoned these operations years ago."[6]

Under Mozambican land legislation, families who have occupied and farmed lands for more than a decade, such as those farming the lands of the old Unidade de Namele farm, are supposed to be granted DUATs that prohibit any company or state agency from displacing them from the lands unless it is clearly in the public interest, such as for the construction of hospitals, schools or highways.

However, local farmer leaders say that Mozaco has already evicted 1,500 farmers to make way for their operations. The organization ADECRU calculates that several thousand more will lose their lands if the company is allowed to expand to 20,000 hectares.[7] And access to land is only part of what's at stake for the communities: Mozaco no doubt chose the area because it is situated between two important rivers, the Malema and the Nataleia, where 4,500 families live and farm. These families now risk losing access to their lands and the water they need to farm and survive.

During the 2012-2013 season, Mozaco cultivated soybeans on about 200 hectares. In its second season, the company expanded to 400 hectares. Ten families lost their homes in the process, and were paid compensation ranging from MT3,000 ($90) to MT10,000 ($300). The local church of Santa Lucia was also destroyed and 1,500 farmers had their access to lands in the area taken away, without any compensation and in complete violation of the land law.[8]

JFS Holding is 100-per-cent owned by the Ferreira dos Santos family of Portugal. They have a long history of involvement in agriculture in Mozambique and JFS is today the largest cotton company in the country. The majority owner of Mozaco, however, is Rioforte Investments, with 60 per cent of the company's shares.[9]

Rioforte is a Luxembourg-headquartered company that was set up in 2009 to hold the non-financial assets of Grupo Espírito Santo - a Portuguese financial dynasty with deep political connections, which is currently embroiled in perhaps the worst economic scandal ever to hit Portugal.

In May 2014, the Banco de Portugal issued an audit questioning the financial stability and transparency of Grupo Espírito Santo's main company, the Banco Espírito Santo. This was followed in August by a controversial €4.5 billion rescue of Banco Espírito Santo, with backing from the EU.

As part of the rescue package, Banco Espírito Santo was divided into two banks: one composed of the "good assets" and one composed of the "toxic assets". These toxic assets consisted mainly of the bank's investments in the largely unregulated and unaudited companies of the Grupo Espírito Santo.

Investigators in at least six countries – Portugal, Switzerland, Venezuela, Panama, Luxembourg and Angola – are reported to be poring over bank documents, transfers and deals, trying to determine what tricks the Grupo Espírito Santo may have used to keep itself afloat.[10]

It appears that Rioforte's assets, including its farms, were dumped in the "toxic" pile. Beyond its Mozaco farming operation in Mozambique, Rioforte owns three soybean and cattle farms in Paraguay covering 135,000 hectares through its subsidiary Paraguay Agricultural Corporation (Payco), and three eucalyptus and food crop farms in Brazil through two other subsidiaries, covering 32,000 hectares.[11]

It is not clear what will now happen with Mozaco and Rioforte's other farms. In July 2014, Rioforte Investments, with nearly $3 billion in debts, requested protection from its creditors in a Luxembourg court – a request that was granted. But in October 2014, the Commercial Court of Luxembourg reversed its decision and ruled that the BES Group subsidiary was to be liquidated and the resulting funds used to pay off its creditors. Grupo Espírito Santo's efforts to appeal the decision were denied.

Banco Espírito Santo also owns 49 per cent of Moza Banco, the fourth-largest private bank in Mozambique. It is not yet clear what the collapse of the Espírito Santo empire will mean for this bank, which is 51-per-cent owned by a consortium of Mozambican investors, led by the former governor of the Bank of Mozambique Prakash Ratilal and in which former President Guebuza is said to have shares.[12]

Box 2 – Profile: AgroMoz

The profile of AgroMoz speaks volumes about the transformation under way in the Nacala Corridor. This company, a partnership involving the richest man in Portugal, the former president of Mozambique and one of the largest land holders in Brazil, has set up operations in the heart of Nacala's soybean-producing zone.

In 2012, AgroMoz representatives arrived at the administrative post of Lioma, hastily arranged for rights to lands with some government authorities and proceeded to evict more than 1,000 peasants from Wakhua village from their lands.[13]

"The process started in 2012 and, at the time, we were told that the AgroMoz project was to deal with an area estimated at only around 200 hectares to begin with, a plot to test the productivity of several seed varieties such as soybeans, corn and beans," says Agostinho Mocernea, the secretary of the village of Nakarari. But the company quickly expanded.[14]

In the 2013/2014 season, AgroMoz cultivated 2,100 hectares, planting soybeans on 1,700 hectares and rice on the other 400 hectares. The company says its intention is to reach 12,000 hectares.[15] The evicted farmers received minimal compensation, ranging from MT2,000 to MT6,500 ($65-$200). One of the farmers, Fernando Quinakhala, a father of five children, says AgroMoz evicted him from a 3.5 hectare plot of land that he and his ancestors farmed. The company determined that he was entitled to MT6,500 in compensation, but Quinakhala says the compensation was nowhere near what the land is worth to him and his family. "I didn't take the money because it was quite insignificant," he says.[16]

According to another farmer from Wakhua, Mariana Narocori, a mother of three children, when the procedure for the granting of land began, she was summoned to participate in a meeting advertised by the local leader, where it was announced that the lands would be given to AgroMoz.[17]

"I was forced to sign a document whose contents I didn't have access to, and I received only MT4,500($155)," says Narocori. "One week later, a bulldozer arrived and demolished my house and destroyed the crops. I was homeless and had to move to the town of Nakarari where I was assigned a plot of land on which I built my house and farm to survive."[18] Her story shows how the displacement of people from Wakhua puts pressure on lands in other areas and creates risks of more land conflicts.

AgroMoz has not fulfilled its promise to the community to construct a clinic and a school. It is, however, already badly affecting the health of the local people. Last season the company commenced aerial spraying of pesticides on its soybean crops.

"In the 2013/2014 agricultural campaign, a group of AgroMoz workers came to tell us that during the spraying, carried out by a small plane, people had to leave their homes as a way to prevent possible harm caused by the chemical," says Mocernea. After a few days, almost all the residents began to suffer from the flu and their crops died.[19]

Despite the opposition from local people and the destructive impacts of the company's acitivities thus far, the Mozambican government granted AgroMoz a DUAT for 9,000 hectares in Lioma. At the time, Armando Guebuza, one of the investors in AgroMoz, was still president of the country.

AgroMoz is reported to be a joint venture between the Grupo Américo Amorim of Portugal, a holding company of Portugal's richest man Américo Amorim, and Intelec, which the US embassy has described as "an investment vehicle for President Guebuza".[20] The Pinesso Group of Brazil, which operates farms on more than 180,000 hectares in Brazil and 22,000 hectares in Sudan, handles the agricultural operations, but it is not clear if they also own a share in the company.

Information from company registry documents and employee websites suggests that AgroMoz is in fact part of AGS Moçambique, SA, a Mozambican company owned by two Portuguese subsidiaries of Grupo Amorim (Solfim SGPS and Sotomar — Empreendimentos Industriais e Imobiliários, SA) and ESF Participaçoes, a subsidiary of ESF Investimentos, which is owned by Intelec and SF Holdings, both of them headed by Guebuza's main business partner, Salimo Abdula.

Full report, including more profiles of the major landgrabbers: https://www.grain.org/e/5137

1 Banco de Moçambique (2013), Relatorio Annual, http://www.bancomoc.mz/Files/CDI/RelatorioAnual2013.pdf

2 Vunhanhe, J. and Adriano, V. (2014) "Segurança Alimentar e Nutricional em Moçambique: um longo caminho por trilhar, Fevereiro, www.r1.ufrrj.br/.../Estudo_de_caso_SAN_em_Mocambique

3 *Idem.*

4 The Mitsui company website states: "Mitsui has the potential to work with Brazil-based SLC Agricola to produce in Portuguese-speaking countries like Angola and Mozambique, should Africa open up to large-scale agriculture", https://www.mitsui.com/jp/en/business/challenge/1201987_1856.html

5 Rioforte Annual Report 2013

6 Interview with a community member affected by the Mozaco project (Malema, July 2014)

7 Ntauz, C. (2014) "Peasants accuse presidential candidates of marginalising small scale agriculture", ADECRU, 6 October, http://farmlandgrab.org/24165

8 Lei de Terra e o decreto (2012) "Regulamento sobre o Processo de Reassentamento Resultante de Actividades Económicas" Point 2, Article 24 of Decree No 31 states that resettlement without proper authorization of the competent authorities is subject to a fine of between MT2-5 million ($60,000-$150,000) and the implementation of an unauthorized resettlement plan is subject to a fine equal to 10% of the budget of the overall project.

9 Rioforte (2013) "Consolidated Financial Statements", http://www.rioforte.com/empresas/ESCOM1/documentos/Rioforte%20Consolidated%20Financial%20Statements%202013.pdf

10 Ellis, E. (2014) "Downfall of a dynasty: The last days of Ricardo Salgado and Banco Espírito Santo", *Euromoney*, 14 October, http://ericellis.com/downfall-of-a-dynasty-the-last-days-of-ricardo-salgado-and-banco-espirito-santo/

11 Rioforte (2013) *op.cit.*

12 This claim is made by the US Charge d'Affaires Todd Chapman in a cable released by Wikileaks, http://leakwire.org/cables/cable/09MAPUTO797.html

13 Paulino, J. (2014) "Mozambique: More than 1,000 people displaced from their lands in Lioma", @*Verdade*, 24 October, http://farmlandgrab.org/24164

14 *Ibid.*

15 Rungo, J. (2014) "Agromoz introduz arroz de sequeiro", *Jornal Domingo*, 6 April, http://www.jornaldomingo.co.mz/index.php/economia/3199-agromoz-introduz-arroz-de-sequeiro

16 Paulino, J. (2014) *op.cit.*

17 The information on the situation in Wakhua comes from Júlio Paulino, *op.cit.*

18 Paulino, J. (2014) *op.cit.*

19 Paulino, J. (2014) *op.cit.*

20 See: http://leakwire.org/cables/cable/09MAPUTO797.html

2.7 Socially responsible farmland investment: a growing trap

In 2012, GRAIN published a report arguing that "regulation" is a misguided approach to stopping the scourge of land grabbing.[1] By regulation, we mean efforts to impose constraints, norms, rules or standards on land deals in order to make them less harmful to people and the environment. Far from turning farmland acquisitions into win-win propositions, we showed how the development of "standards" was simply generating a whole new industry to accredit "responsible" land deals, thus absolving them of the "land grab" label. We argued that these approaches were superficial at best, and primarily aimed at securing social acceptance for the expansion of an agricultural model that only benefits a small number of elites.

What has happened since 2012? A lot more of the same. Those most actively pushing for norms, guidelines, protocols and regulations on land grabbing appear to be the corporations themselves. They need such frameworks to allow them to continue doing business and making money without too many people protesting. And governments and intergovernmental agencies are following suit; in the past few years, they've come up with a dizzying array of new guidelines and principles to regulate land grabbing. A wide range of civil society organizations have also become involved in pushing for norms on land grabbing, either by drafting principles, helping broker deals that adhere to certain standards, or trying to use some of these texts or the political space around them as tools for rural communities to assert their rights.

In our experience, so-called responsible farmland investing is generally bad news. At first glance, it may seem like a good idea. Who could argue with a code of ethics intended to guide agribusiness investment? But both politically and in practice, it rarely works to the advantage of local communities. Rather, it creates a mirage of accountability that responds to the needs of investors, donor agencies and politically influential elites. What we are witnessing on the ground with most of these so-called responsible investment schemes is nothing more than a public relations exercise.

Regulating land grabs: corporations move ahead

Due to increasing public scrutiny, companies are under growing pressure to not be branded as land grabbers or linked to deforestation and other negative environmental or social impacts of farmland investments. To avoid consumer boycotts or legal measures that could restrict their operations, they are rushing to generate their own internal norms, or adhere to external ones, that can put a stamp of "responsible investment" on their plantations, farmland funds, shareholdings or supply chains (*See Box 1: How big is socially responsible investing? As big as China, below*). For example:

- The number of signatories to the United Nations Principles for Responsible Investment (UN PRI) rules on farmland doubled between 2011 and 2014, and the UN PRI has now incorporated those rules into its general guidance for investors;[2]
- Peoples Company, a big land investment facilitator in the US, has produced an in-depth white paper on responsible farmland investment;[3]
- Credit Suisse and other financial companies have issued guidance on responsible agribusiness investing for private equity firms active in emerging economies, with an emphasis on farmland acquisitions;[4] and
- Individual corporations such as Illovo Sugar and Nestlé are publishing their own internal codes of conduct on farmland investing.[5]

Box 1: How big is socially responsible investing? As big as China

Whether you call it "ethical" investing, "sustainable" investing, "impact" investing, "environmental, social and governance" or "guided investing", meeting certain standards as a way of doing business has moved from being trendy to being a dominant approach. In the US, at the end of 2014, "socially responsible investing" or SRI represented $6.6 trillion or 18% of the whole pool of professionally managed investments of $36 trillion.[6] That reflects a growth rate of 76% since 2012. In Europe, SRI represents 11%, or €2 trillion, of the whole pool of professionally managed assets of €18.2 trillion.[7] That reflects a growth rate of 23% for investments in sustainability, 92% for investment in exclusions (e.g. no nuclear or

---

no GMOs) and 132% for impact investing (investments that generate positive social returns in addition to financial gains) since 2011. In Australia and New Zealand, SRI represents a whopping 50% of all professionally managed investments or AU\$630 billion.[8] For these three markets alone, following the finance industry's own definition of "social responsibility," we are talking about nearly \$10 trillion. That is the GDP of China.

What does all of this talk about responsible farmland investing look like on the ground? Stefania Bracco of the UN Food and Agriculture Organization (FAO) tried to quantify it.[9] She took the Land Matrix database of large-scale land deals in Africa and assessed how many of those follow some standard presented as "responsible investment".[10] The results are sobering. Just one quarter of the land deals were made by companies participating in a certified (third-party validated) socially responsible investment (SRI) scheme. In the specific case of biofuels, one third of the projects had no connection to social responsibility; while for another 20 per cent of projects there was no information available about their SRI status. Similarly, a recent UNCTAD-World Bank study of large-scale agricultural investments looked at 39 established projects in Africa and Asia and found that less than one third (30%) were affiliated with a certified SRI scheme.[11] This means that the majority of farmland deals either proclaim to adhere to standards of corporate social responsibility without being subject to scrutiny, or fall outside any SRI frame.

Some big international civil society groups in the meantime have taken another approach, trying to get global food manufacturers like Unilever, Coca Cola, Pepsi and Nestlé to adhere to certain standards and then giving them public recognition for it. This has been described by a high-level meeting of governments and corporations as a process where companies "coerce" their suppliers to conform to guidelines of responsible business conduct.[12]

While it is always good for corporations to clean up bad practices, internal industry surveys reveal that the primary motivation driving companies to adhere to standards on land investment is the risk to their reputations.[13] In other words, their goal is to avoid the land grab label. It is true that since 2008, public pressure has, in some cases, succeeded in getting companies to pull out of land deals and projects. Evidence from the ground, however, makes it clear that corporate actions to reduce "reputational risk" are rarely synonymous with communities keeping control of their lands.

Governments offer up more guidelines

Governments, mainly from industrialized countries, have also ramped up their efforts to facilitate responsible farmland investments. They do this primarily by trying to translate the Voluntary Guidelines (VGs) for the Responsible Governance of Tenure of Land, Fisheries, and Forests in the Context of National Food Security, adopted by the Committee on World Food Security (CFS) in May 2012, into national legislation. The European Union (EU) is now pushing the VGs in Africa through at least two independent programs affecting 21 countries.[14] In addition, the G8 New Alliance on Food Security and Nutrition – a set of funder-driven agribusiness projects in Africa, many of which involve large-scale land acquisitions – adopted its own internal guidance for responsible land deals and encourages corporations participating in the New Alliance to put them into practice. Individual donor governments, such as France, the UK and the US, have also developed standards and guidelines that "their" corporations and development co-operation agencies are supposed to comply with (yet seldom do). Finally, the African Union has produced its own guiding principles on large-scale land investments in Africa through the Land Policy Initiative (LPI).[15]

Meanwhile, intergovernmental agencies and multi-sector groups are drawing up numerous new tools for farmland investors to use to prove their compliance with standards of good corporate behavior.

Box 2: Guidelines galore

- Agence française du développement, together with the Comité Technique Foncier et Développement, has its own operational guide to due diligence (2015) for French investors
- CFS Principles for Responsible Investment in Agriculture and Food Systems (October 2014)
- CFS Voluntary Guidelines on the Responsible Governance of Tenure of Land, Fisheries and Forests in the Context of National Food Security

---

(May 2012) + Operationalising the Voluntary Guidelines on the Responsible Governance of Tenure: a Technical Guide for Investors (September 2015)

- UK DFID is developing how-to guides on "Responsible Investment in Land and Property" and Landesa (connected to Bill Gates) is the group assigned to produce them
- FAO/IFAD/UNCTAD/World Bank Principles for Responsible Agricultural Investment (2009)
- The G8 New Alliance has adopted (June 2015) an Analytical Framework for Responsible Land-Based Agricultural Investments which harmonizes donors' operating principles and aligns them with the CFS Voluntary Guidelines and the LPI's Guiding Principles
- IFC Performance Standards and IFC Voluntary Agro-Commodity Standards: Good Practice Handbook Roadmap to Sustainability (2013)
- Interlaken Group, a collaborative involving major transnational corporations, governments, UN agencies and NGOs has released a land and forest rights guide on how investors can implement the VGs (2015)
- The Land Policy Initiative (African Union, African Development Bank and UN Economic Commission for Africa) Guiding Principles on Large-Scale Land-Based Investments in Africa (2014)
- The OECD, together with the FAO, due diligence guidance (2015)
- The Roundtable for Responsible Soy certification standards; the Roundtable for Sustainable Biofuels guidelines for land rights; the Roundtable for Sustainable Palm Oil Principles and Criteria for the Production of Sustainable Palm Oil; Bonsucro production standard for sugar; and a number of other standards for responsible cotton, coffee, cocoa, etc.
- UN Guiding Principles on Business and Human Rights
- UN PRI Principles for Responsible Investment in Farmland as of September 2014
- USAID Operational Guidelines for Responsible Land-Based Investment (March 2015)
- World Bank safeguards and standards, currently being revised (as of July 2015)

For the internet links to these documents, please see the online version of this article: https://www.grain.org/e/5294

Civil society: making progress or losing out?

A number of civil society organizations and social movements have also been promoting responsible investment as a matter of strategy. For instance, many groups have been pushing for the implementation of the VGs at the national and regional levels. Although acknowledging that the text is not perfect (it does not condemn land grabbing, for example) they see it as providing political support to communities' land rights. Some have been doing this through UN- or government-led initiatives, such as an FAO program in Senegal, in which many national groups are participating, or the African Union's Land Policy Initiative, in which regional networks are engaging or considering engaging. In other cases, international networks such as FIAN, IPC and ActionAid are running their own programs to promote and implement the tenure guidelines at country level. These efforts target not only Africa, but also aim to get the VGs incorporated into national law everywhere, including Europe, Latin America and Asia.

Thus far, Guatemala is the only country that has integrated the VGs into a national land policy framework.[16,17] The country has one of the most unequal landholding structures in the world, with 60 per cent of its farmland devoted to large-scale plantations for export. The new Integrated Rural Development Law is supposed to address this historic injustice and strengthen the rights of peasants and indigenous peoples to their lands. Yet it makes no mention of land redistribution and provides no tangible support to peasant production, upholding instead the existing market-based system which has only accelerated land concentration in the countryside.[18]

Some organizations, such as Friends of the Earth, Fern, Global Witness and ActionAid, have undertaken a different tack, working to get the EU to reform its financial legislation to include the screening of investments for land grab-bing-related criteria. The idea is to ensure that financial institutions like banks and pension funds are required to engage in landgrab-free lending, spending and investing and to back that up with sanctions. But the prospect of creating strong anti-land grab regulations of this sort is quite far off. Given the current political context, in which few European governments are interested in reining in finance, many more years of heavy campaigning would be required before significant headway could be achieved.

Another CSO-supported endeavor over the last few years was the negotiation of a set of principles for responsible agricultural investment ("RAI") within the Committee on Food Security at the FAO.[19] The RAI principles were meant to go a step further than the Voluntary Guidelines on land tenure and establish agreed norms of behavior for corporate investment in food and agriculture more broadly. Many civil society groups and networks supported and participated in the negotiation of these principles. For La Vía Campesina and others, the idea was to assert the importance of small food producers as investors and clear the way for their needs and interests to take center stage. Instead, however, this view got sidelined by other interests and the final text has been denounced by many CSOs who participated in the negotiations.

An analysis by the Transnational Institute highlights some of the main problems with the CFS RAI: human rights are subordinated to trade rules; free, prior and informed consent of indigenous people is included, but subject to reservations; the principles envision a weak regulatory role for the state, leaving current power imbalances intact; farmers' rights are coupled with the interests of seed companies; and although civil society fought hard for the inclusion of agro-ecology, it appears only alongside references to the corporate-friendly term "sustainable intensification".[20] In the French Land and Development Technical Committee's scathing assessment, the CFS RAI does little more than condone the World Bank's RAI.[21]

RAI gone wrong

In practice, "responsible" agricultural investment frameworks seem to be backfiring – or at least proving irrelevant.

Feronia

Take the case of Canadian company Feronia, which has 120,000 hectares of concessions in the Democratic Republic of Congo for oil palm plantations and large-scale cereal farming. The company is 80-per-cent owned by the UK government's CDC Group, together with development finance agencies of France, Spain and the US. Feronia and its shareholders all have policies and standards addressing environmental and social issues, working conditions and financial integrity. Moreover, Feronia has a "zero-tolerance" policy on corruption. The Spanish government shareholder is prohibited from investing in any activity that involves "unacceptable risk to contribute to or be complicit in human rights violations, corruption or negative social or environmental impacts", while the CDC's participation requires that Feronia's operations not be the subject of any environmental, social or land claims. The African Agriculture Fund, through which French and Spanish state stakes in the project are channelled, has its own Code of Conduct for Land Acquisition and Use, but refuses to make it public. Beyond these internal rules, Feronia and its shareholders have also collectively committed to adhere to standards managed by the World Bank, the International Finance Corporation, the Organization for Economic Co-operation and Development and the International Labor Organization.

Yet Feronia is in serious breach of these standards. Its plantations were acquired without the consent of local communities and in murky circumstances involving multimillion-dollar payouts to a close aide of DRC President Joseph Kabila. In testimonies to GRAIN and RIAO-RDC, local community leaders describe horrific working conditions that violate national labor laws. Villagers cannot use any of the lands within the concession areas for agriculture or livestock, even the abandoned areas, and they are beaten, whipped and arrested by company guards if they are caught with oil palm nuts gathered from the plantation area. So far, the only practice that Feronia has had to carry out as a condition for financing is to conduct a CDC-imposed environmental and social assessment of its palm oil operations.[22]

The RSPO

Consider the Roundtable for Sustainable Palm Oil (RSPO), set up in 2004 under the leadership of the World Wildlife Fund (WWF) and several of the world's largest food and plantation companies. For the companies, the RSPO was a means to protect the growth in consumption of a hugely profitable commodity from growing criticism about massive deforestation, land conflicts and labor exploitation. Some of the NGOs that initially signed up to the RSPO saw it as an opportunity to address the power imbalance between communities and workers on the one hand and powerful companies and complicit governments on the other.

On paper, the RSPO has some strong language around free, prior and informed consent (FPIC). Most importantly, it has a grievance mechanism that communities and workers can use to defend themselves against companies that fail to meet the criteria. But as one of the RSPO's longstanding NGO members admits, "industry non-compliance with the RSPO standard is ubiquitous".[23]

In Liberia, for instance, RSPO member Golden Agri-Resources, one of the largest oil-palm plantation companies in the world, signed a 225,000 hectare land deal with the Liberian government. The Forest People's Programme, as part of an FAO project to put the voluntary guidelines into practice, conducted a review of the deal and found no trace of FPIC, despite Liberian land laws requiring it and Golden Agri-Resources' stated commitment to it. The affected communities took their complaints to the RSPO but to no avail. The company is "still manifestly failing to comply with many relevant RSPO, legal and other best practice standards", notes the Forest People's Programme. "Most worrying of all is the picture that emerges of companies whose current business model fundamentally undermines any prospect of their project's community engagement achieving FPIC compliance."[24]

In Malaysia, another RSPO member, Felda Global Ventures, was recently exposed for human rights and labor violations. Felda, which has amassed 700,000 hectares of oil-palm plantations in both Malaysia and Indonesia, is no small player. Its buyers include the US corporation Cargill, which provides oil to Procter & Gamble and Nestlé. A July 2015 investigation by the *Wall Street Journal* showed how workers are being trafficked into Felda's labor force, paid below minimum wage, poorly housed and abused.[25] "They buy and sell us like cattle," one of the Bangladeshi workers said, referring to the contractors who organize Felda's workforce, 85 per cent of whom are migrants.

Transparency is the number one principle of responsible investing for RSPO certification, as well as for most schemes promoting responsible investment, yet there are numerous examples of how transparency fails in practice. In Gabon, the Singaporean oil-palm giant Olam put together a public-private partnership with the Ali Bongo regime to cultivate 50,000 hectares in order to produce RSPO-certified palm oil. Already, 20,000 hectares of forest have been cleared. According to local researcher Franck Ndjmbi, Olam was supposed to conduct a feasibility study before cutting the forest, but no such study was produced.[26]

Box 3: IPOP – Land grabbing in disguise

Another key pillar of responsible farmland investing is "sustainability". The oil-palm sector again provides a strong example of why this principle is so problematic in practice. In September 2014, the four companies that control 80% of Indonesia's palm oil production signed the Indonesia Palm Oil Pledge (IPOP) with the backing of the US State Department.[27] The pledge is supposedly meant to help stop deforestation for the production of palm oil. But in return for staying out of primary forests, the companies receive a license from the Indonesian government to grab lands elsewhere, which typically means land being used by communities (so-called "degraded" land). To implement IPOP, the companies are calling on the government to "codify the elements of the pledge into law". Specifically, they want Indonesia's policy on land swaps to be amended so that companies can more easily "shift their operations from forested to degraded land".[28]

"We're serious about producing palm sustainably, but we need strong regulations that enable us to protect high-carbon stock forests and high-conservation areas," said Cargill Indonesia CEO Jean-Louis Guillot. The government, however, is crying foul, claiming that the companies are trying to dictate law. "The pledge already breaches the State Constitution. We lose our sovereignty because we are controlled [by the pledge]. Our authority is being taken over by the private sector," said San Afri Awang, a representative of the Environment and Forestry Ministry.[29]

For many, IPOP is land grabbing in disguise. In the name of responsible investing, the oil-palm giants gain access to even more lands and lock in that access through new legal instruments.

Other examples

Reports from other experiences keep pouring in. In Nigeria, new on-the-ground research from Friends of the Earth shows how Wilmar, the world's top palm oil producer, is breaching its own responsible investment standards in Cross River State where it currently cultivates 30,000 hectares and has plans for hundreds of thousands.[30] The abuses committed range from non-compliance with the company's obligations on FPIC to large-scale environmental destruction. In Laos PDR, Chinese investors who recently got a 10,000-hectare land concession to produce rice in Chapassek Province were expected to comply with the government's new "fair" investment model. This model requires that the farmers be joint stakeholders in the project through their labor or land contributions. In reality, colleagues report, the villagers received no share of the project's earnings nor was their consent sought before their lands were taken.[31]

In other cases, outright conflicts over the implementation of investment standards have broken out. In Tanzania, for instance, communities and civil society organizations have raised serious complaints about Eco-Energy, a Swedish-led joint venture to produce biofuels. The project is supported by the African Development Bank, the International Fund for Agricultural Development and the Swedish International Development Authority. It involves the production of sugarcane on 20,000 hectares. Almost 1,300 people displaced by the project claim that the company has violated Performance Standard No. 5 of the International Finance Corporation on involuntary resettlement.[32] But the company rejects their claims, calling them "invaders".[33]

Box 4: Road to nowhere?

In 2014, PepsiCo, one of the world's leading industrial food manufacturers, agreed to implement a series of conditions put forward by Oxfam in its "Behind the Brands" campaign in order for its products to be labelled "land–grab–free". The company began implementing the conditions and then produced an audit report to demonstrate how it was faring in its sugar supply chain in Brazil. Oxfam America found the company's way of accounting for its performance lacking in several regards and is now urging PepsiCo to improve.[34] While Oxfam's campaign was certainly well intentioned, this example illustrates what regulating land deals can lead to: international NGOs auditing the audits of transnational corporations that are trying to meet the criteria of the NGOs. Will this really stop the problem of land grabbing on the ground? Is this where we should be putting our energy?

Even in the United States, new reports detail how subcontractors for the Hancock Agricultural Investment Group – one of the country's biggest farmland investment brokers, owned by Canada's largest insurance company, ManuLife – systematically violated domestic labor and safety laws.[35] News of this only emerged because of legal action taken by the workers, something that few farmworkers are able to do. The case shows how the very structure of corporate land deals – in which, for example, an investor places money in a fund that pays a manager who pays a contractor who pays a subcontractor who engages in illicit activity – allows the system to evade responsibility. It also raises serious questions about how the Canadian and US governments can push responsible standards abroad when they are not able to enforce them at home. Indeed, the US food industry – like its counterparts from Australia to Great Britain – is rife with evidence of human trafficking, slavery and other deplorable conditions.

Where to draw the line
The bottom line is that the push for so-called responsible investment in agriculture is not stopping land grabbing. In our view, the reasons for this are structural and stubborn. They include:

- The voluntary nature of all these rules and guidelines fails to create legitimacy and therefore cannot lead to change. Who decides what "responsible" is? What guarantees are there that investors will comply?
- Companies know that they cannot be held to higher standards than national laws. If a country's laws do not recognize community land rights or other rights as "legitimate", they cannot be made to uphold them.
- There is a political choice to be made between promoting agribusiness and promoting community-led farming and food systems. Those who argue that they are compatible or that they must be made compatible are the elites. For the communities who have to give up their lands and livelihoods to make way for large-scale agribusiness projects, compatibility is a myth.

This brings us to the question: what works? What has succeeded in or contributed to stopping land grabs in the last few years? Where should civil society focus its efforts? We see that two things have helped the most. First, there is no doubt about it, political pressure works. What companies call "hype" – media work, public scrutiny, campaigns, mobilizations, inquiries, resistance and direct action – actually drives investors out and away. We have seen this with Gulf State investors and European companies operating in Africa. We have seen projects stopped or scaled back in Cameroon, Tanzania and Madagascar. Communities relentlessly demanding their lands back have also had some success in Sierra Leone (Addax), Cameroon (Herakles), Tanzania (Serengeti) and elsewhere. Of course, this is not overnight work. But it is essential, and in desperate need of serious support.

Second, exposing land grabs for what they really are – violent and devastating, and in many cases unlawful – can work too. Land deals have flopped or been terminated due to corruption, disrespect for human rights, tax evasion and the like. Legal inquiries in Colombia revealed a massive level of fraud being committed by Cargill in its land acquisitions there, leading to legislative change thanks to a bold and progressive political bloc in the congress. Mounting evidence about wrongdoings committed by Indian investor Karuturi in Africa have brought the company under scrutiny and into the courts; Karuturi is now struggling to stay afloat.[36] In Senegal, investigative work by civil society revealed the shady origin and structure of the Senhuile-Senethanol project, which led to its director being fired and jailed (although the project persists).[37] Important work by Global Witness to expose the role of Vietnamese "rubber barons" – and their supporters at Deutsche Bank and the World Bank – grabbing land in Cambodia and Laos for rubber production with impunity, is triggering changes.[38] The point is that shedding light on the criminality that often underpins land deals might be a more useful approach than making the investments more responsible.

Of course, there is a need for diverse strategies and tactics. But for civil society groups, it is politically important to draw the line instead of trying to make land investments nicer, tamer, more inclusive, more sustainable and less abusive. Land grabbing, even under the best practices, is not compatible with food sovereignty, human rights and community wellbeing. It must be exposed for what it is and stopped as a matter of urgency.

This article is available online at: https://www.grain.org/e/5294

1 GRAIN (2012) "Responsible farmland investing? Current efforts to regulate land grabs will make things worse", August, https://www.grain.org/e/4564.

2 See: PRI commodities work stream, http://www.unpri.org/areas-of-work/implementation-support/commodities/

3 Peoples Company (2015), "White paper on socially responsible farmland investing shows benefits of new practices", 4 May, http://peoplescompany.com/blog/2015/white-paper-on-socially-responsible-farmland-investing-shows-benefits-of-new-practices

4 See Credit Suisse et al. (2015) "Private equity and emerging markets agribusiness: Building value through sustainability", 13 May, http://asria.org/publications/private-equity-and-emerging-markets-agribusiness-building-value-through-sustainability/

5 For Illovo, see http://www.illovosugar.co.za/Group-Governance/Group-Guidelines-on-Land-and-Land-Rights. For Nestlé, see Chris Arsenault, C. (2015) "Large food firms back voluntary plan to stop land grabbing", *Reuters*, 17 August, http://mobile.reuters.com/article/idUSL5N10S2Z620150817

6 The Forum for Sustainable and Responsible Investment (2014) "Report on US Sustainable, Responsible and Impact Investing Trends 2014", http://www.ussif.org/Files/Publications/SIF_Trends_14.F.ES.pdf

7 Eurosif (2014) "European SRI study 2014", http://www.eurosif.org/publication/view/european-sri-study-2014/

8 Responsible Investment Association Australasia (2015) "Responsible investment on the rise again to 50% of the investment industry," 11 August, http://responsibleinvestment.org/wp-content/uploads/2015/08/RIAA-RI-On-The-Rise-Again-To-50-Of-The-Investment-Industry-FINAL-MR.pdf

9 Bracco, S. (2015) "Large-scale land acquisitions in Africa: Exploring players, roles and responsibilities," unpublished manuscript, 2015 (copy on file with GRAIN).

10 The Land Matrix is a database of large-scale land acquisitions maintained by a group of mainly academic institutions: http://www.landmatrix.org/.

11 The World Bank (2014) "The practice of responsible investment principles in larger-scale agricultural investments: Implications for corporate performance and impact on local communities", http://unctad.org/en/PublicationsLibrary/wb_unctad_2014_en.pdf

12 Dutch Embassy in Riyadh (2015) "Roundtable on international Responsible Agricultural Investment: proposal for a trilateral approach", Report of a high-level meeting, 20 January, Residence of the Netherlands Ambassador, Diplomatic Quarter, Riyadh, http://www.agroberichtenbuitenland.nl/golfstaten/wp-content/uploads/sites/26/2015/03/imgcq683tsm.pdf

13 Schanzenbaecher, B. and Allen, J. (2015) "Responsible investments in agriculture, in practice: Results and conclusions from a case study review", EBG Capital AG, http://aldenimpact.com/wp-content/uploads/2015/03/RESPONSIBLE-INVESTMENTS-IN-AGRICULTURE-IN-PRACTICE.pdf

14 See AFSA and GRAIN (2015), "Land and seed laws under attack: Who is pushing changes in Africa?", January, pp 8–9, https://www.grain.org/e/5121

15 UNECA (2014) "Guiding Principles on Large Scale Land Based Investments in Africa", African Union, AfDB and UNECA, http://www.uneca.org/publications/guiding-principles-large-scale-land-based-investments-africa

16 FAO (2014) "Integration of the Voluntary Guidelines into land policy in Guatemala", *VGs Newsletter*, November, http://www.fao.org/nr/tenure/whats-new/november-2014-newsletter/en/

17 Itzamná, O. (2015) "¿Desarrollo Rural Integral para qué y para quiénes?" *Rebelión*, 1 June http://www.rebelion.org/noticia.php?id=193983

18 *Ibid.*

19 The principles for responsible agricultural investment developed at the CFS are sometimes identified by the lower case acronym "rai" to distinguish them from the principles for responsible investment developed by the World Bank, FAO, IFAD and UNCTAD ("RAI" or "PRAI")

20 Kay, S. (2015) "Political brief on the principles on responsible investment in agriculture and food systems", TNI, March, https://www.tni.org/en/briefing/political-brief-principles-responsible-investment-agriculture-and-foodsystems

21 Comité technique – Foncier & développement (2014) "État des lieux des cadres normatifs et des directives volontaires concernant le foncier", octobre, http://www.foncier-developpement.fr/wp-content/uploads/Etat-des-lieux-des-cadres-normatifs1.pdf

22 For more information on Feronia, see GRAIN and RIAO (2015) "Agro-colonialism in the Congo", June, https://www.grain.org/e/5220

23 Forest Peoples Programme (2015) "To make palm oil 'sustainable' local communities must be in charge", 14 May, http://www.theecologist.org/campaigning/2856781/to_make_palm_oil_sustainable_local_communities_must_be_in_charge.html

24 Forest Peoples Programme (2015) "Hollow Promises: An FPIC assessment of Golden Veroleum and Golden Agri-Resource's palm oil project in Liberia", 15 April, http://www.forestpeoples.org/topics/agribusiness/publication/2015/hollow-promises-fpic-assessment-golden-veroleum-and-golden-agri

25 Al-Mahmood, S.Z. (2015) "Palm-oil migrant workers tell of abuses on Malaysian plantations", *Wall Street Journal*, 26 July, http://www.wsj.com/articles/palm-oil-migrant-workers-tell-of-abuses-on-malaysian-plantations-1437933321

26 VOA Afrique (2015) "Le Gabon veut devenir le premier producteur africain d'huile de palme", 17 August, http://www.voaafrique.com/content/le-gabon-veut-devenir-le-premier-producteur-africain-d-huile-de-palme/2920679.html

27 The four companies are Golden Agri Resources, Wilmar, Asian Agri and Cargill. See Eco-Business (2015) "Palm giants ask Indonesian gov't to clear path toward sustainability", 4 May, http://www.eco-business.com/news/palm-giants-ask-indonesian-govt-to-clear-path-toward-sustainability/. They have since been joined by Musim Mas, the fifth-largest oil palm producer in the country, and the Chamber of Commerce and Industry of Indonesia

28 Eco-Business (2015) *op.cit.*

29 Jong, H.N. (2015) "Gov't opposes zero-deforestation pledge by palm oil firms", *The Jakarta Post*, 29 August, http://www.thejakartapost.com/news/2015/08/29/govt-opposes-zero-deforestation-pledge-palm-oil-firms.html

30 Friends of the Earth (2105) "Exploitation and empty promises: Wilmar's Nigerian landgrab", June, http://www.foe.org/news/news-releases/2015-07-worlds-largest-palm-oil-trader-comes-under-scrutiny

31 Darren Daley of GAPE, based in Champassak, communication with GRAIN

32 See the campaign launched by ActionAid in March 2015: http://www.actionaid.org/publications/take-action-stop-ecoenergys-land-grab

33 Eco-Energy (2015) "Land grabbing definition perpetrated in BEE Report, 31 March, http://www.ecoenergy.co.tz/fileadmin/user_upload/AA_Report_Response.pdf

34 Oxfam America (2015) "PepsiCo publishes audit on land rights in Brazil", 8 July, http://politicsofpoverty.oxfamamerica.org/2015/07/pepsico-publishes-audit-on-land-rights-in-brazil/

35 See Oakland Institute (2014) "Down on the farm", http://www.oaklandinstitute.org/sites/
    oaklandinstitute.org/files/OI_Report_Down_on_the_Farm.pdf
36 See the "Karuturi" section of farmlandgrab.org, http://farmlandgrab.org/cat/show/348
37 Collectif du Ndiaël and Re:Common (2015) "Senegal land grab on the verge of implosion",
    24 July, http://farmlandgrab.org/25156
38 Global Witness (2013), "Rubber Barons", 13 May, https://www.globalwitness.org/
    campaigns/land-deals/rubberbarons/

3
The struggle for seeds

3.1 Seed laws that criminalize farmers

Seeds are one of the irreplaceable pillars of food production. Farmers all over the world have been acutely aware of this throughout the centuries. It is one of the most universal and basic understandings that all farmers share. Except in those cases where they have suffered external aggressions or extreme circumstances, almost all farming communities know how to save, store and share seeds. Millions of families and farming communities have worked to create hundreds of crops and thousands of varieties of these crops. The regular exchange of seeds among communities and peoples has allowed crops to adapt to different conditions, climates and topographies. This is what has allowed farming to spread and grow and feed the world with a diversified diet.

But seeds have also been the basis of productive, social and cultural processes that have given rural people the resolute ability to maintain some degree of autonomy and to refuse to be completely controlled by big business and big money. From the point of view of corporate interests that are striving to take control of land, farming, food and the huge market that these factors represent, this independence is an obstacle.

Ever since the Green Revolution, corporations have deployed a range of strategies to get this control: agricultural research and extension programs, the development of global commodity chains, and the massive expansion of export agriculture and agribusiness. Most farmers and indigenous peoples have resisted and continue to resist this takeover in different ways.

Today, the corporate sector is trying to stamp out this rebellion through a global legal offensive. Ever since the establishment of the World Trade Organization (WTO), and almost without exception, all countries of the world have passed laws giving corporations ownership over life forms. Whether through patents or so-called plant breeders' rights or plant variety protection laws, it is now possible to privatize micro-organisms, genes, cells, plants, seeds and animals.

Social movements worldwide, especially peasant farmers' organizations, have resisted and mobilized to prevent such laws being passed. In many parts of the world, the resistance continues and can even count some victories. To strengthen this movement, it is very important that as many people as possible, especially in the villages and rural communities that are most affected, understand these laws, their impacts and objectives, as well as the capacity of social movements to replace them with laws that protect peasants' rights.

Today's seed laws promoted by the industry are characterized by the following:

- They are constantly evolving and becoming more aggressive. Through new waves of political and economic pressure – especially through so-called free trade agreements, bilateral investment treaties and regional integration initiatives – all the "soft" forms of ownership rights over seeds were hardened and are rapidly being made more restrictive. Seed laws and plant variety rights are being revised again and again to adapt to the new demands of the seed and biotechnology industry.
- Laws that grant property rights over seeds have been reinforced by other regulations, which are supposed to ensure seed quality, market transparency, prevention of counterfeits, etc. These regulations include seed certification, marketing and sanitary rules. By means of these regulations, it becomes mandatory, for instance, for farmers to purchase or use only commercial seeds tailored for industrial farming. Or the regulations make it a crime to give seeds to your son or exchange them with a neighbor. As a result, seed fairs and exchanges – a growing form of resistance to control over seeds – are becoming illegal in more and more countries.
- While strengthening privatization, these laws disregard basic principles of justice and freedom and directly violate the Universal Declaration of Human Rights. Such seed laws impose the rule that anyone accused of not respecting property rights over seeds is assumed to be guilty, thus violating the principle that people are innocent until proven guilty. In some cases, measures can be taken against accused wrongdoers without their being informed of the charges. These seed laws are even making it an obligation to report alleged transgressors; they are legalizing searches and seizures of seeds on grounds of mere suspicion (even without a warrant) and allowing private agencies to conduct such checks.
- These laws are being drafted in vague, incomprehensible and contradictory language, leaving much room for interpretation. In most cases, the laws move through legislative chambers in secrecy or by means of international agreements that cannot be debated nationally or locally.

Once the misinformation and secrecy used to push the laws through have been countered by information campaigns and mobilization on the part of social organizations, experience shows that people do not want these laws. Indeed, most people reject the idea that a company can take ownership of a plant variety and prohibit farmers from reproducing their seeds. They find it completely absurd. People also generally do not agree that the work that farmers do to feed the world should suddenly become a crime. Wherever resistance has been strong enough, the legal plunder embodied in these laws has been stopped.

Experience also shows that those who want to privatize, monopolize and control seeds on behalf of large transnational corporations have no limits. There is no possibility to negotiate, make concessions, or reach common agreements on this in a way that would allow the different interests to co-exist peacefully. The corporate agenda is to make it impossible for farmers to save seeds and to make them dependent on purchased seeds.

Similarly, experience shows that it is possible to resist and dismantle these attacks. But doing so requires informative tools that can be widely shared, in order to blow away the smoke of false promises and nice words, so that people can see what really lies behind seed laws.

How seed laws make farmers' seed illegal

The displacement of peasant seeds is a process that has been gaining ground and speed around the world over several decades. In the 20th century, when plant breeding and seed production became activities separate from farming itself, peasant varieties were gradually replaced by industrial varieties. In Europe and North America, this happened over several decades, spurred by new technologies such as the development of hybrids. In Asia, Africa and Latin America, it took off after the 1960s, when so-called development programs pushed "high-yielding" crops and the use of chemical inputs (the so-called Green Revolution). In the past 20 years, we have witnessed a new situation in which an aggressive wave of seed laws is being unleashed, often in the name of liberalizing trade, with the purpose of stopping nearly all activities carried out by farmers with their seeds.

Farmers who produce and exchange their own seeds within their own community or with neighboring communities are not in need of laws to govern their actions. The collective rights to use community seeds, which are often oral, are established and respected enough within each community for such use to be regulated. But once the seeds are commercialized on a large scale by companies who produce them with unknown methods and in unknown locations, often beyond national borders, then laws become necessary in order to combat fraud, counterfeiting, bad quality seeds that do not germinate or that carry diseases, as well as to regulate GMOs. Laws are also necessary to protect local seeds and the social and cultural systems which guarantee the survival of the population's chosen systems of food production. These laws for "prevention of commercial fraud" and the protection of food sovereignty represent a conquest on the part of rural organizations. Unfortunately, however, once the pressure of mobilization by popular organizations and farmers weakens, most of these laws are rewritten by the industry in order to promote their own industrial "improved" seeds, and to ban farm seeds.

The term "seed laws" often refers to intellectual property rules, such as patent laws or plant variety protection legislation. But, in fact, there are many other laws pertaining to seeds, including those that regulate trade and investments; regulations related to the health of plants; certification and so-called "good agricultural practices" related to marketing; or so-called biosafety regulations (*See Box 1: Types of seed laws promoted by industry, on page 134*). As a whole, these laws often result in peasant seeds being decreed illegal, branded as inadequate, and treated as a source of risk to be eliminated.

The new seed laws are a reflection of the increasing power of the food and agriculture industries. Until the 1970s, new types of crop varieties were developed and distributed by state-run companies, small seed houses, and government research stations. Since then, we have witnessed a massive process of large companies taking over smaller ones and public programs giving way to the private sector. Today, just 10 companies account for 55 per cent of the global seed market. And the lobbying power of giants such as Monsanto, Dow or Syngenta is enormous. As a result, they have managed to impose restrictive measures giving them monopoly control.

Trade and investment agreements are a weapon of choice to impose seed laws where they did not exist before, or to make existing laws more favorable to transnational corporations. The end goal is clear: to prevent farmers from saving seeds so that they buy corporate seeds on the market instead. And in that process, to get governments to pull out of plant breeding and seed production. In Africa, farmers' seeds represent 80-90 per cent of what is planted each season. In Asia and Latin America, they account for up to 80 per cent. So, from the perspective of an agribusiness CEO, there is still a huge market out there to create and capture. Even in Europe, where industrial seeds already dominate farming, corporations continue lobbying for stronger enforcement of existing regulations in order to eliminate pockets of resistance and to restrict farmers' abilities to reuse industrial seeds. In the cases when these laws are enforced the results are very repressive: farmers' seeds are confiscated and destroyed; farmers are targeted and under surveillance; and some face criminal charges and jail sentences for simply continuing to work within their peasant systems and for using their own seeds.

At the same time, almost everywhere that we look, the power of the industry is also being contested. Challenging this power takes on many different forms, including: organizing and mass mobilizations; countering the false propaganda that these seed laws are necessary or are in the interest of the people; media work; education in schools and places of worship; street theater; civil disobedience in defiance of unfair laws; and, most importantly, the daily work of continuing to develop peasant and small-scale farming systems. These systems include not only the native or local seeds and breeds, but also the land, territories, and rural peoples' cultures and ways of life. Experience shows that when this counter-force to defend peasant seeds is strong, then institutional challenges in the courts or in parliaments can force the suspension of bad laws – or at least call them into question. Given the power and interests that are at stake, overturning these seed laws is not achieved in a single battle. Rather, it is a continuous struggle in defense of peasant agriculture and food sovereignty as a whole.

Box 1: Types of seed laws promoted by industry

Marketing laws are the oldest and most widespread type of regulations affecting seeds. They define the criteria that must be met in order for seeds to be put on the market. As such, they are often justified as a means of protecting farmers, as consumers of seeds, in order to ensure that they are only offered good seeds – both in terms of physical quality (germination rate, purity, etc.) and of the variety in question (genetic potential). But whose criteria are used? In the countries that have adopted the system of "compulsory catalogue", seeds are allowed on the market only if they belong to a variety responding to three critical requirements: they must be "distinct", "uniform" and "stable" (DUS criteria). This means that all plants grown from a batch of seed will be the same, and that their characteristics will last over time. Peasant varieties do not fit these criteria, because they are diverse and evolving. Marketing laws also typically require that your variety present a "value for cultivation and use", usually referring to its yield under mono-cropping cultivation dependent on a large amount of chemical fertilizers. Another problem is how marketing is defined. Under many countries' seed laws, the definition of marketing is not restricted to monetary sales alone. Marketing can include free exchange, bartering, transfer of seeds within networks or even just giving seeds as gifts.

Intellectual property laws applied to seeds are regulations that recognize a person or an entity, most often a seed company, as the exclusive owner of seeds having specific characteristics. The owner then has the legal right to prevent others from using, producing, exchanging or selling them. The justification for this is to give companies a temporary monopoly so that they can collect a return on their investment without facing competition. But there are huge problems involved.

There are two main types of intellectual property systems for seeds: patents and Plant Variety Protection (PVP). The US started to allow patents on plants in the 1930s, when flower breeders demanded a kind of copyright on their "creations"; they wanted to stop others from "stealing" and making money from their flowers. Plant patents are very strong rights: no-one can produce, reproduce, exchange, sell or even use the patented plant for research without the owners' authorization. To use patented seeds, farmers must make a payment to the owner of the patent. Farmers who buy patented seeds are also obliged to agree to a set of conditions: that they will not reuse seed from their harvest for the following season; that they will not experiment with the seeds; that they will not sell or give them to anyone else. The Monsanto Company even asks farmers to spy on their neighbors and report to the police anyone who is doing these things with Monsanto seeds. Today, patenting is standard for GMOs.

Plant Variety Protection is a kind of patent developed in Europe specifically for plant breeders. It is accompanied by the same DUS criteria as those required by the catalogue and it initially granted fewer powers than a patent. In 1961, European states created the Union for the Protection of New Plant Varieties (UPOV), which harmonizes rules through the UPOV Convention, which has been revised several times. The UPOV gives breeders the right over their commercial varieties to prevent anyone else from producing seeds for commercial purposes. However, other breeders can use "protected" (or privatized) materials for breeding programs. In the first decades of UPOV's existence, farmers were still free to save and reuse their seeds from protected varieties. However, with the revision of the UPOV Convention in 1991, protection of plant varieties extends to prohibit the agricultural production of the protected variety, including harvesting and the post-harvest produce. Under UPOV 91, farmers are no longer allowed to reuse seeds of privatized varieties – except in rare cases and upon payment. If farmers infringe the regulation or are suspected of infringement, they can have their houses searched without warrant, their crops, harvests and processed products seized and destroyed, and they could be sent to jail for years. The UPOV 91 also makes it much easier for seed companies to privatize farmers' own farm-produced seeds and to ban the use of local varieties.

Trade and investment agreements are a tool used by corporations to force governments to adopt policies promoting corporate rights over seeds. For example, almost all countries of the world are members of the WTO, which has an agreement on trade-related aspects of intellectual property rights (TRIPS). The TRIPS agreement requires countries either to provide some form of plant variety protection or to face trade sanctions. In addition, many countries have been bullied into joining UPOV 91 – through bilateral free trade agreements, development aid, etc.

Trade agreements such as those required by the WTO, and FTAs, set market rules that supposedly aim to prohibit discrimination but may also give agribusiness preferred access to certain markets. As a result, governments may no longer be able to implement procurement programs under which state authorities buy seeds from local farmers. The rationale is that by restraining competition, local procurement requirements put transnational companies at a trade disadvantage. These are harsh conditions that give preference to corporations rather than to the welfare of farmers or consumers.

Bilateral investment treaties, pushed by countries such as the US and members of the European Union, also contain a rule that intellectual property on seeds is a form of foreign investment that must be protected in the same way as an oil well or car factory. Thus, if such investments are expropriated or nationalized, or if the expected profits from them are jeopardized, then a US or EU seed company can sue the country in which the investment is located in an international court (investor-state dispute settlement).

Plant health and biosafety laws can also limit farmers' use of and access to their seeds. Such laws are intended to prevent health or environmental hazards that can arise from seeds, including contamination through GMOs, and can, in that sense, be useful. Plant health regulations, for instance, are aimed at preventing the spread of diseases via seeds that are produced in one location and exported to another. The problem lies in the fact that these laws actually serve to protect the interests of industry. For example, sometimes small-scale exchanges of seeds among farmers are prohibited, or their seeds are confiscated and destroyed, because farmers are held to the same standards as transnational corporations, which sell seeds in far greater amounts and to more distant locations – with a corresponding increase in the chance of spreading disease. Under such laws, farmers' seeds may be viewed as a potential risk or hazard while industry seeds are hailed as the only safe ones, even though they play a huge role in spreading disease and contamination.

Similarly, biosafety laws often have the opposite effect of what they were intended to do. Instead of setting up barriers to the entry and spread of GMOs, which by their very nature are hazardous, they create a legal framework to manage risks and therefore facilitate the acceptance and spread of transgenic seeds. For example, biosafety laws often lay out formal procedures for planting GMOs that result in standards making these procedures legal without their being any safer. Such laws can also force farmers who do not want GMO and who produce their own seeds to have all their seeds analyzed in order to guarantee the absence of GMO, which they obviously are unable to do, thus obliging them to buy industry-sold GM seeds. In other instances, these laws make it much easier to import or export GM crops, since the countries involved have the necessary legal mechanisms set up to oversee the crops. In yet other cases, such as that of Europe, there are good biosafety laws in place which have preventive measures to stop the cultivation or import of GMOs, but these laws are under fire as the seed industry sees them as barriers to trade.

It should be noted that United Nations agencies such as the UN Food and Agriculture Organization and the UN Conference on Trade and Development or the World Intellectual Property Organization are today important proponents of all of the above laws. They draft model laws and train governments in how to implement them.

Box 2: Six action points

Seeds are under attack everywhere. Under corporate pressure, laws in many countries increasingly put limitations on what farmers can do with their seeds and with the seeds they buy. Seed saving, a thousand-year-old practice that forms the basis of farming, is fast becoming criminalized. What can we do about this?

Defend farmers' own seed systems

Farmers' fields are the first line of defense against bad seed laws. This means organizing to rescue, collect, maintain, develop, share and use local farmers' seeds. It is very important that women and young people are all involved. You can start a project with neighbors or local associations, talk to market or street vendors, get schools or your workplace involved, etc. Seed fairs and visits to farms and gardens are an important part of this work.

Stop "Monsanto laws"

Law proposals criminalizing farmers are easier to fight against before they become written into law. If public opinion is against them, they become more costly for governments to push through. Organize street protests, make videos, talk to the media, organize direct actions...

Join forces with other farmers

In many other countries, farmers are fighting very similar laws. Learning from them and their experiences, good and bad, can be very helpful. Even if we have different strategies, we can build common fronts against the seed industry and the governments acting in their interest.

Build alliances with other movements

Broad alliances can be built when people understand that seeds affect everyone's wellbeing, not only farmers'. The struggle for seeds can be integrated into farmers' wider struggles, since there is no food sovereignty without seed sovereignty. Seed struggles can also be important parts of larger fights, such as campaigns and actions against free trade agreements, austerity measures, new patent or internet regimes, climate change, land laws, etc.

Undo the propaganda

Seed companies and governments present seed laws as protecting consumers, ensuring quality seeds, raising yields and feeding the hungry. We need to debunk these myths and show that the agriculture they are promoting is toxic and generates hunger. These laws are only meant to extract wealth.

Try to get positive laws

In some cases, it may be possible to obtain favorable laws, programs or tools that protect farmers' seed systems. Think of GM-free zones, laws rejecting patents on life or programs that promote local varieties and farmers' seeds. In other cases, such laws or legal efforts may exclude people, divide communities, entangle farmers in legal bureaucracies, create contradictions or be a waste of time.

The great climate robbery

This chapter is extracted from the joint Via Campesina – GRAIN publication with the same name. The full version, with numerous country cases and other materials, can be downloaded from: https://www.grain.org/e/5142

3.2 Trade deals and farmers' seeds

What could be more routine than saving seeds from one season to the next? After all, that is how we grow crops on our farms and in our gardens. Yet from Guatemala to Ghana, from Mozambique to Malaysia, this basic practice is being turned into a criminal offense, so that half a dozen large, transnational corporations can turn seeds into private property and make money from them.

But people are fighting back and in several countries popular mobilizations are already forcing governments to put seed privatization plans on hold.

Trade agreements have become a tool of choice for governments, working with corporate lobbies, to push new rules to restrict farmers' rights to work with seeds. Until some years ago, the most important of these was the World Trade Organization's (WTO) agreement on Trade-Related Aspects of Intellectual Property Rights (TRIPS). Adopted in 1994, TRIPS was, and still is, the first international treaty to establish global standards for "intellectual property" rights over seeds.[1] The goal is to ensure that companies like Monsanto or Syngenta, which spend money on plant breeding and genetic engineering, can control what happens to the seeds they produce by preventing farmers from re-using them – in much the same way as Hollywood or Microsoft try to stop people from copying and sharing films or software by putting legal and technological locks on them.

But seeds are not software. The very notion of "patenting life" is hugely contested. For this reason, the WTO agreement was a kind of global compromise between governments. It says that countries may exclude plants and animals (other than micro-organisms) from their patent laws, but they must provide some form of intellectual property protection over plant varieties, without specifying how to do that.

Trade agreements negotiated outside the WTO, especially those initiated by powerful economies of the Global North, tend to go much further. They often require signatory countries to patent plants or animals, or to follow the rules of the Geneva-based Union for the Protection of New Plant Varieties (UPOV) that provide patent-like rights over crop varieties. Whether in the form of patent laws or UPOV, these rules generally make it illegal for farmers to save, exchange, sell or modify seeds they save from so-called protected varieties.[2] In fact, in 1991 the UPOV convention was modified to give even stronger monopoly powers to agribusiness companies at the expense of small and indigenous farming communities. This 1991 version of UPOV now gets widely promoted through trade deals.

Onslaught of FTAs

The North America Free Trade Agreement – signed by Mexico, Canada and the US, at about the same time TRIPS was being finalized – was one of the first trade deals negotiated outside the multilateral arena to carry with it the tighter seed privatization noose. It obliged Mexico to join the UPOV club of countries giving exclusive rights to seed companies to stop farmers from recycling and reusing corporate seeds. This set a precedent for all US bilateral trade agreements that followed, while the European Union, the European Free Trade Association (EFTA) and Japan also jumped on the same idea.[3]

A non-stop process of diplomatic and financial pressure to get countries to privatize seeds "through the back door" (these trade deals are negotiated in secret) has been going on since then. The stakes are high for the seed industry. Globally, just 10 companies control 55 per cent of the commercial seed market.[4]

But for these corporations, that market share is still not enough. Across Asia, Africa and Latin America, up to 80 per cent of the seeds farmers use are farm-saved seeds, whether from their own farms or from neighbors or nearby communities. In these unconquered territories, the agribusiness giants want to replace seed saving with seed markets and take control of those markets. To facilitate this, they demand legal protections from governments to create and enforce corporate monopoly rights on seeds. This is where free trade agreements come in as a perfect vehicle to force countries to change their laws.

Latest trends

For the past 15 years, GRAIN has been tracking how trade deals signed outside the multilateral system are coercing countries to adopt the industry's wish-list of intellectual property rights for seeds, and ratchet up global standards in that process. A recent update of our dataset shows that this trend is not letting up. In fact, there are worrisome signs on the horizon:

- The most important recent gains for Monsanto, Dupont, Limagrain and Syngenta – the world's top seed companies – have come from new trade deals accepted by Latin American states. In 2006, the US (home to Monsanto and Dupont) closed major deals with Peru and Colombia, forcing both countries to adopt UPOV 1991. The EFTA states (home to Syngenta) did the same in 2008 and the EU (home to Limagrain) in 2012.[5] In Central America, a similar pattern occurred. The US secured a very powerful Central America Free Trade Agreement in 2007, forcing all countries to adhere to UPOV 1991. The EFTA did the same in 2014.

- An important step towards stronger proprietary seed markets was recently taken in Africa. After 10 years of talks, Economic Partnership Agreements (EPAs) were concluded between the EU and sub-Saharan African states in 2014. Most of them "only" liberalize trade in goods for now, but also contain a commitment to negotiate common intellectual property standards with Brussels. The expectation is that those standards will be based on what the Caribbean states already agreed to in their 2008 EPA: an obligation to at least consider joining UPOV. This is significant because until now African states have been under no obligation to adopt UPOV as a standard, and actually tried to come up with their own systems of plant variety protection.[6] And while it's true that African entities like the anglophone African Regional Intellectual Property Organization (ARIPO) and the francophone African Intellectual Property Organization (OAPI) are already joining UPOV, under the EU trade deals the countries themselves would be the ones to join. Further towards the horizon, Africa is harmonizing within itself as its subregional trade blocs merge and unite to form a single continental free trade zone, supposedly by 2017. This is expected to bring with it an internal harmonization of intellectual property laws across the continent, likely tightening the noose even further.

- The Trans-Pacific Partnership (TPP) agreement will seriously undermine farmers' rights to control seeds in Asia and the Pacific. This is because the US-driven agreement, signed in February 2016, requires UPOV 1991 to be applied in all member countries. It also says that inventions 'derived from plants' – which for the biotechnology industry means GM seeds – are patentable. We don't yet know whether these ratcheting demands will also appear in the Transatlantic Trade and Investment Partnership (TTIP) currently being negotiated between the US and the EU, because the text remains inaccessible to the public.

- While the extent of what has to be privatized expands, so do the penalties for disrespecting these norms. Under numerous FTAs, countries like the US require that farmers who infringe these new intellectual property rights on seeds face punishment under criminal law instead of civil law. In some cases, like the recently concluded EU-Canada Comprehensive Economic and Trade Agreement (CETA), the mere suspicion of infringement could see a farmer's assets seized or their bank accounts frozen.[7]

Big battles heating up

The good news is that social movements are not taking this sitting down. They are becoming very active, vocal, bold and organized about this. In 2013, Colombians from all walks of life were shaken when they saw first-hand how US and European FTAs could result in their own government violently destroying tonnes of seeds saved by farmers who did not know what the new rules were. The outrage, which was expressed in the midst of a massive national agrarian strike, was so strong that the government actually agreed to suspend the law temporarily and re-examine the issue directly with farmers' representatives.[8]

In 2014, it was Guatemala's turn to be rocked when the general public realized that the government was pushing through the adoption of UPOV 1991 without proper debate. Because of trade deals like CAFTA,[9] people were furious that indigenous communities were not consulted as is required, especially when the purpose of the law – ultimately – is to replace indigenous seeds with commercial seeds from foreign companies like Monsanto or Syngenta. After months of pressure, the government backed down and repealed the law.[10] But – as in Colombia – this retreat is only temporary while other measures are considered. In yet other parts of Latin America, such as Chile and Argentina, new laws to implement UPOV 91, often dubbed "Monsanto Laws", are also being intensely and successfully resisted by social movements. In Africa too, waves of public protest have been reported against the plant variety protection regimes that countries are adopting. In Ghana, a vibrant campaign is under way to stop the government from adopting UPOV 1991 legislation.[11] Elsewhere, civil society networks like the broad-based Alliance for Food Sovereignty in Africa are filing appeals to stop ARIPO from adopting UPOV-based legislation and joining the union.[12]

Corporate interest groups have pushed too far trying to privatize what people consider a commons. This is not limited to seeds. The same process has been going on with land, minerals, hydrocarbons, water, knowledge, the internet, even important micro-organisms, like avian flu a few years ago or the Ebola virus today. People are fighting back to stop these things falling under the exclusive control of a few corporations or defense ministries. A good way to take part in this battle is to join the campaigns to stop important new trade deals like TTIP, CETA, TPP and the EPAs – and to get old ones like the US and European deals with Mexico, Central America, Colombia or Chile rescinded. It is in the process of trade deals where many of these rules get written and where they should be erased.

For the original version of this paper: https://www.grain.org/e/5070
For a closer look at the status of trade agreements that impose seed privatization, download
GRAIN's November 2014 dataset: http://www.grain.org/attachments/3247/download

1 "Intellectual property" is a government enforced monopoly right. It serves to ensure that people pay for the right to use something for a certain period of time, so that whoever invented it can recoup his or her investment. "Plant variety" means seeds which will grow into a specific kind of plant with specific characteristics.

2 Under the UPOV system, farmers can sometimes save seeds from protected varieties to use them again. It depends on which version of the UPOV Convention a country signs and whether the government exercises this option. Sometimes it is restricted to farmers' replanting the seeds on their own farm or to only certain crops or to payment of a license. Under the patent system, it is simply illegal to use patented seeds without paying for them – even if a bird drops them onto your field!

3 EFTA is composed of Iceland, Liechtenstein, Norway and Switzerland

4 ETC Group (2008) "Who owns nature?", http://www.etcgroup.org/content/who-owns-nature

5 Ecuador is also now negotiating with the EU, based on the text signed with Colombia and Peru

6 For example, the Organization of African Unity drafted its own model law on plant variety protection based on community rights

7 See National Farmers' Union (2014) "CETA + Bill C-18 = too much power for seed companies", June, http://www.nfu.ca/sites/www.nfu.ca/files/CETA%20and%20C-18%20Fact%20Sheet%20-%20June%202014.pdf

8 GRAIN (2013) "Colombia farmers' uprising puts the spotlight on seeds", September, http://www.grain.org/e/4779

9 Perhaps not very visible to the public eye was the 2013 EFTA-Central America FTA, which makes the same demands as CAFTA.

10 See EFE (2014) "Guatemala repeals plant breeder rights law", 5 September, http://www.bilaterals.org/?guatemala-repeals-plant-breeder

11 See the websites of Food Sovereignty Ghana, http://foodsovereigntyghana.org/ and Panafricanist International, http://www.panafricanistinternational.org/

12 Alliance for Food Sovereignty in Africa (2014), "AFSA appeals to ARIPO, AU and UNECA for protection of farmers' rights & right to food", 2 July http://www.acbio.org.za/index.php/media/64-media-releases/462-alliance-for-food-sovereignty-in-africa-media-briefing-afsa-appeals-to-aripo-au-and-uneca-for-protection-of-farmers-rights-a-right-to-food

3.3 GMOs: Feeding or fooling the world?

They want us to believe that GMOs will feed the world; that they are more productive; that they will eliminate the use of agrichemicals; that they can coexist with other crops, and that they are perfectly safe for humans and the environment. False in every case, we'll show how easy it is to debunk these myths. All it takes is a dispassionate, objective look at 20 years of commercial planting of genetically engineered (GE) crops and the research that supposedly backs it up. The conclusion is clear: GMOs are part of the problem, not part of the solution.

GE crops will end world hunger
FALSE.

Genetically engineered crops have nothing to do with ending world hunger, no matter how much GE spokespeople like to expound on this topic. Three comments give the lie to their claim:

- FAO data clearly show that the world produces plenty of food to feed everyone, year after year. Yet hunger is still with us. That's because hunger is not primarily a question of productivity, but of access to arable land and resources. Put bluntly: Hunger is caused by poverty and exclusion.
- Today's commercial GE crops weren't designed to fight hunger in the first place. They aren't even mainly for human consumption. Practically the entire area planted to GE crops consists of soybeans, corn, rapeseed, and cotton. The first three of these are used almost exclusively to make cattle feed, car fuel, and industrial oils for the United States and Europe, while cotton goes into clothing.
- More damning, there appears to be an iniquitous cause-and-effect relationship between GE crops and rural hunger. In countries like Brazil and Argentina, gigantic "green deserts" of corn and soybeans invade peasants' land, depriving them – or outright robbing them – of their means of subsistence. The consequence is hunger, abject poverty and agrotoxin poisoning for rural people. The truth is that GE crops are edging out food on millions of hectares of fertile farmland.

In the year GMO seeds were first planted, 800 million people worldwide were hungry. Today, with millions of hectares of GMOs in production, 1 billion are hungry. When exactly do these crops start "feeding the world"?

GE crops are more productive

FALSE.

Look at the data from the country with the longest experience of GMOs: the United States. In the most extensive and rigorous study, the Union of Concerned Scientists[1] analyzed 20 years of GE crops and concluded that genetically engineered herbicide-tolerant soybeans and corn are no more productive than conventional plants and methods. Furthermore, 86 per cent of the corn productivity increases obtained in the past 20 years were due to conventional methods and practices. Other studies have found GE productivity to be lower than conventional methods.

Crop plants are complex living beings, not Lego blocks. Their productivity is a function of multiple genetic and environmental factors, not some elusive "productivity gene". You can't just flip a genetic switch and turn on high productivity, nor would any responsible genetic engineer make such a claim. Even after all this time, GE methods are quite rudimentary. Proponents of the technology count it a success if they manage to transfer even two or three functional genes into one plant.

The bottom line is that two decades and untold millions of dollars of research have resulted in a grand total of two marketable traits – herbicide tolerance and Bt pest resistance (see below). Neither has anything to do with productivity.

GE crops will eliminate agrichemicals

FALSE.

In fact, it is the reverse. GE crops increase the use of harmful agrichemicals. Industry people try to put this myth over by touting the "Bt gene" from the *Bacillus thuringiensis* bacteria, which produces a toxin lethal to some corn and cotton worms. The plants produce their own pesticide, supposedly obviating the need to spray. But with such large areas planted to Bt monocultures, the worms have quickly developed resistance to Bt; worse, a host of formerly unknown secondary pests now have to be controlled with more chemicals.

The other innovation trumpeted by the "genetically modified corporations" consists of plants that can withstand high doses of herbicides. This allows vast monocultures to be sprayed from the air, year after year on the same site. It's a convenience for industrial farmers, which has abetted the spectacular expansion of soybeans in recent years. Thirty years ago there were no soybeans in Argentina; now they take up half the country's arable land. Concurrently, the amount of the herbicide glyphosate sprayed in Argentina has skyrocketed from 8 million liters in 1995 to over 200 million liters today – a 20-fold increase, all for use in GE soy production.

The same thing is happening in the United States. Herbicide-tolerant GMOs have opened the floodgates, and glyphosate and other herbicides are pouring onto farmers' fields. In 2011, US farmers using this type of GMO sprayed 24 per cent more herbicides than their colleagues planting conventional seeds.[2] Why? For reasons any evolutionary biologist could have predicted: the weeds are evolving chemical resistance. In short, the GE "revolution" is an environmental problem, not a solution.

Farmers can decide for themselves. After all, GMOs can peacefully coexist with other crops

FALSE.

Genetically engineered boosters may claim nobody's forcing farmers to use GMOs, but a pesky little fact of basic biology implicates non-GE farmers against their will. It's called cross-pollination: Plants of the same species interbreed, and sooner or later the genes artificially inserted in the GE crops cross into the conventional crops.

In Canada, the widespread growing of genetically engineered canola has contaminated nearly all the conventional canola and in so doing wiped out organic canola production. Similar contamination has been found in corn crops around the world.

The introduction of GE seed is especially alarming when there is potential for contamination of local varieties. Mexico is the center of origin and diversification of corn. For years now, Mexican indigenous communities have been noticing odd traits appearing in some of their varieties. Various studies confirm that this is because of contamination by GE corn imported from the United States. Now, the Mexican government is proposing to allow transnationals to plant up to 2.4 million ha of GE corn in the country. If this project goes ahead, it will not only be an attack on the food sovereignty of the Mexican people: it will be a threat to the biodiversity of one of the world's most important staple food crops.

In the Spanish state of Aragón, farm and environmental organizations have been complaining since 2005 that more than 40 per cent of organic grain has traces of GE content and can no longer be sold as organic or GMO-free.

What's really perverse about this fake "freedom to farm" argument is that certain transnationals have been forcing farmers to pay for seeds they never planted. In the United States, Monsanto has taken hundreds of farmers to court for supposedly infringing its intellectual property rights. Monsanto detectives roam the countryside like debt collectors, looking for "their genes" in farmers' fields. In many cases, the genes got there because the farmers either purchased contaminated seed or had their own crops contaminated by a neighbor's field. Whatever the case, it's a lucrative strategy that has brought in millions of extra dollars for the corporation. And it has the added benefit of scaring farmers away from buying anything but Monsanto seeds. Sounds a lot more like the "freedom" to do exactly what the transnationals tell you to.

Box 1: Genetic engineering – a stalled science

GE crops are in the hands of very few companies. Monsanto most notoriously, along with Dupont, Syngenta, BASF, Bayer and Dow, dominate GE research and patents, corner 60 per cent of the world seed market and control 76 per cent of the world agrichemical market.

Yet all the profitable "science" owned by these companies comes down to two and only two traits: herbicide tolerance and Bt.

In 2012, 59 per cent of the area planted to commercial GE crops consisted of crops resistant to the herbicide glyphosate, a product originally patented by Monsanto, while 26 per cent consisted of insecticidal Bt crops and 15 per cent consisted of crops carrying both traits.

Two traits. That's all these transnationals have to show for 20 years of research and mega-millions of dollars invested. Some revolution! The real measure of what GE technology has produced is to be found in damaged ecosystems, potential health harms, farmer dependency – and big profits for the companies.

GE crops pose no threat to health and the environment
FALSE.

At the very least, the biosafety of transgenic crops is an open question. Do we really want to entrust our health to an industrial agriculture system in which GE purveyors control food security offices and dictate their own standards? Food sovereignty requires that the people, not the companies, have control over what we eat.

Nevertheless, our plates are now filling up with food items from plants with altered DNA and heavy pesticide loads, and we are told to simply shut up and eat. Concerns have been heightened by a number of credible reports on GMOs and their attendant herbicides:

- The American Academy of Environmental Medicine (AAEM) stated in 2009 that genetically engineered foods "pose a serious health risk". Citing various studies, it concluded that "there is more than a casual association between GE foods and adverse health effects" and that these foods "pose a serious health risk in the areas of toxicology, allergy and immune function, reproductive health, and metabolic, physiologic and genetic health."
- The latest studies by Dr Gilles-Éric Séralini looked at rats fed glyphosate-tolerant GE maize for two years. These rats showed greater and earlier mortality in addition to hormonal effects, mammary tumors in females, and liver and kidney disease.
- A recent study at the University of Leipzig (Germany) found high concentrations of glyphosate, the main ingredient in Roundup, in urine samples from city dwellers – from 5 to 20 times greater than the limit for drinking water.
- Professor Andrés Carrasco of the CONICET-UBA Molecular Embryology Lab at the University of Buenos Aires medical school has unveiled a study showing that glyphosate herbicides cause malformations in frog and chicken embryos at doses much lower than those used in agriculture. The malformations were of a type similar to those observed in human embryos exposed to these herbicides.

Finally, there is the incontrovertible evidence that glyphosate can have a direct impact on human beings, causing miscarriages, illnesses and even death in high enough doses, as explained by Sofía Gatica, the Argentine winner of the latest Goldman prize.[3]

The struggle for seeds

Original: https://www.grain.org/e/472

1 See: http://www.ucsusa.org/food_and_agriculture/our-failing-food-system/
  genetic-engineering/failure-to-yield.html
2 See: http://www.motherjones.com/tom-philpott/2012/10/
  how-gmos-ramped-us-pesticide-use
3 See: http://www.goldmanprize.org/recipient/sofia-gatica/

3.4 Yvapuruvu Declaration: seed laws – resisting dispossession

This declaration was adopted in Yvapuruvu, Paraguay, on 18 October 2013 by Alianza Biodiversidad, Red por una América Latina Libre de Transgénicos, and Vía Campesina World Seeds Campaign

Seeds are the work of peoples and a part of their history. They have been created through collective work, creativity, experimentation and stewardship. Seeds in turn have raised peoples, making possible their specific food-ways and cropping systems and allowing them to share and develop their world views. Seeds are therefore intimately linked to community standards, responsibilities, obligations, and rights. Seeds give us responsibilities that in fact precede our right to use them.

Seeds are the fundamental basis of sustenance. If the whole world now feeds itself from agriculture – enjoying the flavors and forms of food; if agriculture sustains all of humanity, it is because peoples have stewarded them, taken them along on their journeys and allowed them to circulate. And now seeds, as the basis of our sustenance and our existence, are under attack. The purpose of this attack is to put an end to peasant and indigenous agriculture, and especially to independent food production. It is an attempt to foreclose on the future of food sovereignty, turning us into a landless population, good for nothing but cheap labor and dependency. It is an attack mounted in various forms and via a range of mechanisms. We must confront it on every front.

The highest profile aspect of the attack on seeds, and everything they represent, is that of intellectual property. Most commonly this takes the form of what are now called plant breeders' rights or "UPOV laws", but it also includes certification laws, varietal registration and marketing laws. These are laws and regulations that legalize abuse and dispossession.

Specifically:
1. They allow companies to appropriate native seeds.
2. They prohibit and criminalize the use, saving, handling, exchange, and reproduction of native seeds.

3. They allow for the confiscation and destruction of our seeds, crops, and harvests.
4. They force us to accept incursions into our land, storehouses and homes, even with military intervention.
5. They provide for fines and jail terms through legal proceedings that do not even afford us a proper defense, since they presume our guilt.

These laws prevent seeds from travelling and evolving with people. They freeze seeds in time, so that they cannot be transformed and adapted to new environments. They condemn our seeds to death.

This privatization and pillage of what is ours is supported by other provisions now being imposed on us, including: food safety standards; grower and ecosystem certification standards; the misnamed "good agricultural practices"; the latest roll-outs of the "green revolution"; agrochemical packages; phytosanitary standards; environmental services programs; agricultural development and financing programs; the introduction of new technologies, especially transgenics, with the looming threat of terminator crops; integrated production arrangements; contract growing; land-use and zoning plans; associations with big capital, and more.

Corporations, governments, and international agencies have used a series of myths and lies to justify these laws. The first and most shameful is that these laws will give us access to higher-quality industrial seeds. This ignores ample evidence that native seeds are best adapted to actual growing conditions and provide for stable, diverse production of sufficient quantities. It also ignores the fact that far from guaranteeing quality, privatization laws give corporations the power to tie us down to toxic, unreliable seeds.

In reality, this is a war on peoples' sustenance. They want to weaken our capacity to resist. They want us to give up our livelihoods, our lands and our territories, leaving the field free for them to take over our ecosystems, install urban and toxic waste dumps, take possession of our water sources and agrifood system, and expand agribusiness extractivism, agrofuels, mining, deforestation, tree monocultures, dams, tourism, and the occupation of the countryside as the exclusive preserve of the powerful classes.

In the face of these threats, the rural peoples of the world have the duty and the collective and historic right to regain, strengthen and maintain the stewardship of our seeds, our ways of life and our methods of production.

It is a responsibility we have very much already taken on: people's campaigns are springing up all across the continent, and the defense of seeds in the hands of peoples is central to many of them. Today, our organizations and our seeds are fighting off the dispossession that results from all forms of intellectual property or other privatization methods. We will continue to take care of our seeds. We will continue to exchange seeds and knowledge. We will continue to plant our seeds and to teach new generations how to grow them and keep them alive. We will continue to build food sovereignty and to resist agribusiness, along with the whole culture of homogenization, privatization and death that it is trying to impose. We will fight until seed privatization laws in all their guises have disappeared, leaving only a bad memory. It is important for this resistance to continue to broaden and proliferate; we will use various methods of awareness raising and consensus building so as to bring the broadest possible range of sectors into our struggle, because the defense of seeds, and of peasant and indigenous agriculture, is the defense of food and the future of humanity.

In addition to reaffirming our commitments, we greet with joy and pride the many different campaigns being unfurled in our region: the broad mobilization in Mexico to defend corn against the invasion of GMOs and the criminalization of seeds; the Honduran land recovery movement; the Costa Rican campaign that has led to 77 per cent of the country's municipalities being declared GMO-free; the farmers' strike, the quashing of UPOV 91 by the constitutional court, and the resistance against seed confiscation in Colombia; the wide-ranging mobilization against UPOV laws in Chile and Argentina; the mobilizations against agribusiness and soybeans in Brazil, Uruguay, Paraguay, and Argentina, including the blockade of the Monsanto plant in the Malvinas Argentinas district of the city of Córdoba by local residents and members of the Mothers of Ituzaingó; and the demand in Uruguay that the competent authorities take the measures necessary to prevent native maize from being contaminated with transgenes. At the same time we salute the years of campaign and struggle against UPOV that have taken place in Costa Rica since 1999, especially during the resistance to the FTA with the United States (2004-2008).

We repudiate attempts by the Brazilian congress to approve the use of genetic use restriction technologies (GURTs), popularly known as terminator technologies, because of the risks they pose to biodiversity and food sovereignty and because of the violation of indigenous peoples' and peasants' rights that they represent. If Brazil were to go in this direction, it would be unilaterally violating a UN agreement and opening the doors for other countries to be pressured into releasing this technology.

Noting with consternation the situation in Paraguay, where agribusiness has demonstrated its capacity for destruction and domination, we stand in solidarity with the struggle and resistance of the Paraguayan peoples; we commit to accompanying them on their path and to bringing their voices and their example to every corner of our own territories.

We wage our struggle today in an environment that has been depoliticized from the corridors of power on down, in which scorn for all that is rural, peasant or indigenous holds sway; an environment that ignores the knowledge and contributions of rural peoples and communities while presenting big capital, globalization and agribusiness as the only alternatives. For this reason, many of the proposed solutions render us invisible and destroy or ignore the unbreakable bonds between peoples, communities and seeds – the only real basis of any effective possibility of protecting them and safeguarding their future.

We cannot let it be forgotten that the stewardship of seeds is one of humanity's oldest strategies, without which the future is in jeopardy. Seeds are the heritage of peoples; the two have evolved together and are not isolated beings floating in a social void. Seeds are not things, nor are they merchandise or computer programs. They cannot remain in circulation without the stewardship and care of peoples and communities. They are not a resource waiting to be grabbed by the first comer. In other words, there is no such thing as seeds that are free in the abstract. They are free thanks to the peoples and communities who defend, maintain and care for them so that we can enjoy the goods they provide.

Members of Alianza Biodiversidad:

REDES-Amigos de la Tierra, Uruguay. GRAIN, Chile, Argentina and Mexico. Grupo ETC México. Vía Campesina World Seeds Campaign, Chile. Grupo Semillas, Colombia. Acción Ecológica, Ecuador. Red de Coordinación en Biodiversidad, Costa Rica. Acción por la Biodiversidad, Argentina. SOBREVIVENCIA, Amigos de la Tierra Paraguay. Centro Ecológico, Brazil. CLOC-Vía Campesina.

Original: https://www.grain.org/e/4810

4
Controlling
the food system

4.1 Corporations replace peasants in China's new food security agenda

The past few years have been horrible for China's small dairy farmers. Demand for domestic production has slumped because of milk contamination scandals that they had nothing to do with, and milk prices have dropped below the costs of production. Many farmers are choosing to slaughter their cows, but even this desperate act rarely saves them from bankruptcy.[1]

The same cannot be said for China's big dairy companies. These are high times for companies like the New Hope Group and Bright Foods. As they gouge Chinese farmers with low prices for their milk, they are using their profits to establish their own mega dairy farms in China and abroad, especially in New Zealand and Australia, where they can export powdered milk back to China under free trade agreements and market it as "safe".[2]

Liu Yonghao, the Chair of Chinese agribusiness giant New Hope Group, is emblematic of China's new agribusiness leaders. Seizing upon the milk crisis affecting Chinese farmers as an opportunity for his company, he's established an alliance with two of the biggest families in the Australian dairy industry to purchase Australian dairy farms.[3] Their joint venture will begin with the construction of Australia's first 10,000-head dairy farm, followed by two similar sized outfits over the next decade.

This same story has unfolded in respect of other foods in China. The flood of imports of soybeans and palm oil, ushered in with China's accession to the World Trade Organization, fuelled the growth of processed food companies and corporate factory farms for poultry and pigs. Transnational companies such as Cargill, Wilmar and Charoen Pokphand made a killing from these changes, while Chinese farmers were decimated.[4] Many Chinese companies also shut down, but several emerged much stronger, and, with the backing of the state, have been moving aggressively to establish global supply chains and even their own farms both in China and overseas (*See Table 1: China's major food and agribusiness companies, page 159*). The impacts of these changes extend well beyond mainland China and affect peasants around the world.

Table 1 China's major food and agribusiness companies

Company	Activities
Agria Corp	Mainly involved in seeds, fertilizers and rural services. It owns New Zealand's PGG Wrightson, Agrocentro Uruguay and Australia's largest cereal and pulse seeds company, Grainland Moree Pty Ltd.
Chongqing Grain Group (CGG)	State-owned CGG in 2011 set aside $3.4 billion for an overseas expansion that included a 200,000-hectare soybean farm in Brazil, a 130,000-hectare soybean farm in Argentina's Chaco province, and other operations in Canada, Australia, Cambodia and Malaysia. All overseas projects, however, appear to have been put on hold or abandoned for unknown reasons.
CITIC	CITIC is a state-owned financial company active in financing land transfers in China and acquiring farmland abroad, notably for oil palm plantations in Indonesia and a massive 500,000-hectare farm project in Angola. In 2014, Itochu of Japan and Charoen Pokphand of Thailand purchased a combined 20% stake in CITIC.
China National Cereals, Oils and Foodstuffs Corporation Group (COFCO)	COFCO is China's largest food processor, manufacturer and trader. Besides the foodstuff business, COFCO has developed into a diversified conglomerate, from farming, food-processing, finance, warehouse, transportation, port facilities, hotels and real estate. Since acquiring controlling shares in two of the world's largest grain traders, Nidera and Noble, in 2014, COFCO is now one of the world's main grain traders, particularly from the Southern Cone of Latin America.
Heilongjiang Beidahuang	Heilongjiang Beidahuang is China's largest agricultural enterprise, involved in the production of rice, flour, oil, dairy products, pork and potatoes. The state-owned company manages almost 3 million hectares of farmland, 920,000 hectares of forestland and 350,000 hectares of grassland in the province of Heilongjiang. Overseas, it is pursuing farmland in Argentina, Philippines, Australia and, potentially, Russia.

Company	Activities
New Hope Group (NHG)	NHG is the largest private agribusiness enterprise in China, with more than 400 subsidiaries and more than 80,000 employees nationally and abroad. It is the largest producer of feed and largest supplier of dairy, egg and meat products in China. It has set up more than 20 animal feed plants and poultry farms in Southeast Asian countries, including the Philippines, Indonesia, Cambodia and Singapore, and is looking to establish operations on other continents. Its overseas expansion plans are supported by the World Bank, Temasek and several transnational grain traders, notably Mitsui and Marubeni of Japan and ADM of the US.
Shanghai Pengxin	Pengxin is a diversified conglomerate active in real estate, agribusiness, mining, infrastructure and finance. It established a soybean farm in Bolivia in 2005 and has purchased, or is in the process of acquiring, numerous dairy farms in New Zealand and Australia.
WH Group	WH Group of China is the world's largest pork company, with number one positions in China, US and key markets in Europe. It owns Smithfield Foods of the US and major China pork producer Shuanghui Development. In 2013, WH Group recorded sales of more than $20 billion.

"Family farms" displace peasant families

Food security has always been a major concern for China's government. Up until recently that meant ensuring that enough food was produced in China to feed the entire population, and this task fell almost entirely to China's peasant farmers.

Over the past couple of decades, however, the government has shifted its approach to food security, gradually breaking with the old food self-sufficiency policies. Part of the impetus comes from the government's embrace of trade agreements that oblige China to allow imports of certain foods into the country. But the government has also pursued its own domestic policies aimed at shifting food production from peasant farms to larger commercial farms and shifting agriculture extension and procurement from public programs to agribusiness and food corporations.[5]

In 2015, the government put forward a third round of adjustment policy for the agricultural sector in which it says it will enhance previous efforts to reform land holdings, consolidate farms and develop corporate supply chains for inputs (seeds, machinery) and outputs (foods).[6] It will also be providing policies and programs to foster so-called "dragon head enterprises", specialized in vertically integrated supply chains, and to encourage industrial and commercial companies to get directly involved in farming.[7] The effects of this shift in state support from peasants to agribusiness are most advanced with meat production. Twenty years ago, backyard farms supplied China with 80 per cent of its pork; today, it is larger specialized farms and massive factory farms that produce the same amount.[8]

One central pillar of the Chinese state's new agricultural policy is its support for the transfer of land from peasant farms to larger farms, the latter which the government ironically calls "family farms". China's family farms have on average 27 times more farmland than a typical peasant household, and by the end of 2012, there were already around 877,000 such family farms covering 11.7 million hectares of land.[9,10]

Chinese law, however, still prevents peasants from selling their land, so instead the transfers are of "use rights" organized through various schemes, of which perhaps the most important is the land circulation trust. Under this scheme, a company establishes a trust to acquire multiple land-use rights from farmers in a particular area, identify entities interested in the lands, and then arranges for the lands to be leased to these entities. The trust is like a bank where farmers deposit their land rights for the trust to then rent out to much larger farming operations.[11]

The first company to jump into the trust business was the giant state-owned financial company, CITIC. Its founder is China's former vice-president, Rong Yiren – one of Asia's richest men and one of the main politicians responsible for opening up the country's economy to foreign investment. The CITIC's land circulation trusts are done in partnership with the German seed and pesticide corporation Bayer CropSciences, and they integrate Bayer's products into the consolidated farm holdings that they manage.

The CITIC and Bayer's first land trust project in Anhui Province, East China, involves the transfer of 2,100 hectares of farmland from local farmers, who are supposed to receive an average annual payment of 700-800 yuan ($112–128) each. The Chinese government is using the project as a pilot for a nationwide program, launched in 2015, which will register the contractual rights of 200 million rural households over the nation's arable land and pave the way for further transfers. According to China's Ministry of Agriculture, use rights for 25 million hectares – more than one quarter of the total arable land in farmers' hands have already been transferred, either to other farmers or to non-farmers.[12]

Other large companies have followed CITIC and Bayer into the trust business, including China's largest grain trader, COFCO, the US seed company Pioneer, and even China's largest e-commerce merchant, Alibaba.

"Going global": overseas agribusiness investment

In November 2014, the Chinese press reported that CITIC would be investing $5 billion to develop farming operations on 500,000 hectares in Angola. The company has already developed two 10,000-hectare operations in the country and is negotiating with the government for a third farm of 30,000 hectares.[13] The announcement came at the same time that two of Asia's largest food companies, Itochu of Japan and Charoen Pokphand of Thailand, announced a deal under which they would purchase a combined 20 per cent stake in CITIC for $10 billion.[14]

The two deals reveal much about the current direction of China's food security policy. On the one hand, it provides an example of the close integration between the Chinese companies leading the transformation of the Chinese countryside and their foreign counterparts who already dominate the global food trade. On the other hand, it highlights the interest that Chinese companies have in establishing control over the production of food abroad for the export of foods back home.

GRAIN pulled the information available on the farmlandgrab.org website, up to August 2015, to create a database of large-scale overseas land acquisitions by Chinese companies for food production.[15] We identified 61 deals (in process or concluded) in 31 countries covering more than 3.3 million hectares.[16]

What is clear from the data is a focus on the production of certain key foods that China imports (soybeans, palm oil, dairy) or is predicted to be importing much more of (maize, wheat, rice, meat) in the near future. Chinese companies are both looking to establish a presence in the already established centers of export production (US, Argentina, Brazil, Indonesia, Australia, New Zealand) or the areas seen as next frontiers for low-cost production of foods for export (Africa, Cambodia). And there is a tendency to target food production from places that are viewed as having higher "food safety" standards than China, such as New Zealand and Australia, where Chinese companies are interested in producing milk powder for use in infant formula.

In Australia alone, Chinese companies have acquired nearly one million hectares during the past few years, mainly for dairy and beef cattle, and another $700 million worth of deals for rural properties were reported to be under negotiation as of September 2015.[17,18] Several Chinese companies are also in the bidding for Australia's largest landholder, S. Kidman & Co, which owns 11 million hectares of cattle stations.[19] The surge in Chinese interest in Australian farms is tightly connected to the free trade agreement signed between the two countries, which gives protections to Chinese investors and allows for greater exports of Australian foods to China.

The scale of China's new outward investment in global farmland is clearly significant, and is having an impact on local farmers – from Australian farmers being squeezed off their lands because of rising land prices to Mozambican farmers being simply thrown off theirs.[20,21]

More is bound to come. The government is concerned about the country's over-reliance on foreign corporations for food imports and, as a counter strategy, it provides direct support to foster China's own food transnationals and to build the infrastructure and trading logistics that can ensure their access to and control over food exports.[22] As the case of CITIC shows, many of these plans are unfolding through mergers, take-overs and joint ventures with established foreign companies.

The Chinese company that has moved the most aggressively in this direction is the state-owned China National Cereals, Oils and Foodstuffs Corporation Group (COFCO), China's largest processor, manufacturer and trader of food.

"Whatever Chinese consume more of, need more supply of from outside, this is our area," says Frank Ning, the Chair of COFCO.[23]

The COFCO recently acquired controlling stakes in Nidera of the Netherlands and Noble of Singapore, two of the largest traders of grains and oilseeds from the Southern Cone of Latin America. Noble is also an important player in the oil palm trade from Indonesia. The acquisitions were supported by China's sovereign wealth fund, the China Investment Corp (CIC), and made possible by a previous $4.7 million cash injection from the China Development Bank in 2013.[24,25] But COFCO has also sought alliances with foreign financial players to fund its expansion. At home, COFCO sold a major stake in its meat subsidiary, one of the largest in China, to the US private equity firm Kholberg Kravis Roberts (KKR) to raise funds for the construction of mega hog farms. The KKR also has a partnership with leading Chinese dairy producer China Modern Dairy, for the construction of mega dairy farms.[26]

Similarly, when Shuanghui International, a subsidiary of China's WH Group, took over the world's largest pork producer, Smithfield Foods of the US, in 2013, the deal was financed by a $4 billion loan from the Bank of China and funded in part by the Wall Street financial firm Goldman Sachs and Singapore's sovereign wealth fund, Temasek Holdings.[27]

New Hope's overseas expansion plans, co-ordinated from its offices in Singapore, are financed in part through partnerships with the World Bank, Singapore's sovereign wealth fund and several transnational grain traders, notably Mitsui and Marubeni of Japan and ADM of the US.[28,29,30]

To put a number on all this activity: overseas agriculture investment by Chinese companies is reported to have surpassed $43 billion during the past decade.[31]

Chinese agribusiness versus food sovereignty

China has already experienced the largest and fastest rural-to-urban migration in history. It is unlikely, however, that the country's manufacturing sector can continue to absorb this migration, and already there are movements of workers back to the countryside. The current shift to agribusiness means that they will have a much more difficult time surviving back home, as their lands are taken over by larger farms, the markets for their produce are controlled by powerful retailers and food companies, and their environment is polluted by pesticides, chemical fertilizers and waste from factory farms.[32,33]

These developments in rural China coincide with an expansion of Chinese agribusiness and food companies into other countries, where peasants and small farmers are also struggling to maintain access to their lands.

Chinese companies are, of course, not behaving any differently than agribusiness corporations from other countries. In fact, they typically co-operate with foreign food, agribusiness and financial corporations. And this is precisely the problem. China has quickly become a major new source of agribusiness expansion, working directly against the interests of small farmers and local food systems in China and around the world.

Source: www.grain.org/e/5330

1 Smart Agriculture Analytics (2015) "Shandong, Shaanxi and other regions still slaughtering dairy herds; milk cheaper than water", *Dairy Foods*, 27 May, http://www.dairyfoods.com/articles/91176-dairy-news-from-china-milk-prices-fall-farmers-cull-herds-tibetan-dairy-is-worlds-highest-new-bright-dairy-plant-to-process-1500-tons-daily

2 Research and Market (2015) "Research report on China dairy industry, 2015-2019", July, http://www.researchandmarkets.com/research/zbkchn/research_report

3 Kitney, D. (2015) "China's Liu Yonghao New Hope door to Australian dairy", *The Australian*, August, http://www.theaustralian.com.au/business/in-depth/chinas-liu-yonghao-new-hope-door-to-australian-dairy/story-fni2wt8c-1227468594353

4 See GRAIN (2012) "Who will feed China: Agribusiness or its own farmers? Decisions in Beijing echo around the world", August, https://www.grain.org/e/4546

5 Tao, W. (2013) "China wants to deepen overall reforms", *China Daily*, 14 November http://europe.chinadaily.com.cn/opinion/2013-11/14/content_17104733.htm; Ministry of Agriculture of the People's Republic of China, Interpretation of "Guidelines on further adjust and optimise the agricultural structure", February 2015 (in Chinese), http://www.moa.gov.cn/zwllm/zwdt/201502/t20150210_4403210.htm

6 Chinese Government Network (2015) "Li Keqiang: The evolution of the modern agriculture industry", 25 July (in Chinese), http://www.gov.cn/xinwen/2015-07/25/content_2902475.htm

7 Hairong, Y. (2015) "Agrarian capitalization without capitalism?: Capitalist dynamics from above and below in China", *Journal of Agrarian Change*, July, http://www.iss.nl/fileadmin/ASSETS/iss/Research_and_projects/Research_networks/LDPI/CMCP_65-Yan_Chen.pdf

8 GRAIN (2012), *op.cit.*

9 Chinese Ministry of Commerce (2014) "Nongyebu: jiating nongchang 87.7 wan ge (Ministry of Agriculture: Family Farms 877000)", 9 June, http://nc.mofcom.gov.cn/articlexw/xw/gnyw/201406/18719552_1.html

10 Reuters (2014) "China to focus on family farms in drive to commercialise", 27 February, http://www.scmp.com/news/china/article/1436429/china-focus-family-farms-drive-commercialise

11 GRAIN (2015) "Asia's agrarian reform in reverse: laws taking land out of small farmers' hands," 30 April, https://www.grain.org/e/5195

12 GRAIN (2015) "Asia's agrarian reform in reverse: laws taking land out of small farmers' hands", 30 April, https://www.grain.org/fr/article/entries/5195-asia-s-agrarian-reform-in-reverse-laws-taking-land-out-of-small-farmers-hands

13 Wang Bingfei. (2014). "Country rises from ruins to strive", *China Daily Africa*, 9 May, http://africa.chinadaily.com.cn/weekly/2014-05/09/content_17495476.htm. And Xinhua (2015). "Chinese company introducing modern farming technology into Angola", 10 September, http://www.globalpost.com/article/6646473/2015/09/10/chinese-company-introduces-modern-farming-technology-angola

14 Rick Carew and Atsuko Fukase. (2015). "Itochu, CP in deal to take $10.4 billion stake in CITIC", *Wall Street Journal*, 20 January. http://www.wsj.com/articles/itochu-cp-take-10-4-billion-stake-in-citic-1421715748?tesla=y

15 GRAIN (2015) "Overseas large-scale farmland acquisitions for food production made by Chinese companies since 2006", https://www.grain.org/attachments/3638/download

16 Numerous other deals that were announced and are part of GRAIN's 2012 database have since been abandoned and are not included in the table. See, GRAIN (2012) "GRAIN releases

data set with over 400 global land grabs", February, https://www.grain.org/e/4479

17 The $700 million figure includes the October 2015 purchase of Australia's largest dairy farming company. See *The Australian* (2015) "Chinese buy nation's largest dairy, Van Diemen's Land Company", 13 October, http://www.theaustralian.com.au/business/companies/chinese-buy-nations-largest-dairy-van-diemens-land-company/story-fn91v9q3-1227566733334

18 Murray, L. and Cranston, M. (2015) "Chinese investors on buying spree", *Financial Review*, 28 September http://farmlandgrab.org/25345

19 (2015) "China's Donlinks Grain & Oil Co Ltd bids for S.Kidman & Co", *Financial Review*, 1 October, http://farmlandgrab.org/25361

20 See for example, Sunday Night (2015), 'Foreign investors on hunt for Aussie farms", 5 July, https://au.news.yahoo.com/sunday-night/features/a/28634435/foreign-investors-on-hunt-for-aussie-farms/?cmp=fb

21 Ecologist (2013) "China accused of stealth land grab over Mozambique's great rice project", November, http://farmlandgrab.org/22864#sthash.kJs7LdyK.dpuf

22 For example, China signed a deal with Brazil in May 2015 to provide $50 billion towards the construction of a railway link from Brazil's Atlantic coast to the Pacific coast of Peru to reduce the cost of exports to China, particularly for soybeans and other food commodities. See BBC News (2015) "China to invest $50bn in Brazil infrastructure", 15 May, http://www.bbc.co.uk/news/business-32747454

23 Myers, M. (2013) "China's agriculture investment in Latin America", Inter-American Dialogue, 21 November, https://www.bu.edu/pardeeschool/files/2014/12/Margaret-Myers-Lecture1.pdf

24 China Daily (2013) "COFCO gets 30b yuan loan from CDB", 26 February, http://www.chinadaily.com.cn/business//////2013-02/26/content_16259274.htm;

25 Reuters (2015) "China's COFCO, CIC to set up venture to run agricultural businesses", 12 May, http://farmlandgrab.org/24891

26 APK (2015) "KKR backs Cofco drive to ramp up China hog output", 6 June. http://farmlandgrab.org/post/view/23591

27 Ciajing (2013) "Smithfield Foods – Shuanghui International: The biggest Chinese acquisition that isn't", June, http://farmlandgrab.org/22155

28 International Finance Corporation (2015), "IFC partners with New Hope to boost agribusiness in South and Southeast Asia", 9 June, http://ifcextapps.ifc.org/ifcext%5CPressroom%5CIFCPressRoom.nsf%5C0%5CD1F0E4AB4BF5725985257E6000302C48

29 All About Feed (2012) "New Hope and Marubeni join for overseas expansion", 23 January, http://farmlandgrab.org/19927 ;

30 Reuters (2011) "China's New Hope to set up overseas investment fund", 16 November, http://farmlandgrab.org/19617

31 American Enterprise Institute (2015) "China global investment tracker", https://www.aei.org/china-global-investment-tracker/

32 For impacts of factory farms see Schneider, M (2012) "Box 4: Major impacts of the industrialization of meat production in China", in GRAIN, "Who will feed China: Agribusiness or its own farmers? Decisions in Beijing echo around the world", August, https://www.grain.org/e/4546;

33 Another example of the scale of the environmental problems facing Chinese agriculture is soil pollution. See, Dim Sums (2014) "Soil pollution survey finally announced", April, http://dimsums.blogspot.co.id/2014/04/soil-pollution-survey-finally-announced.html

4.2 Defending people's milk in India

"We take care of the cow and the cow takes care of us," says Marayal, a farmer in Thalavady, Tamil Nadu. Her two cows produce up to 10 liters of milk a day, which she sells for 30 to 40 cents per liter.

Across India, there are millions of backyard dairy farmers like Marayal. Each owning just one or two cows, farmers supply millions more families and hundreds of thousands of informal milk parlors and tea stalls across India. These small, unregistered operations prefer to buy milk directly from backyard dairy farmers, who supply fresh milk at the lowest price.

Seventy million rural households in India – well over half of the country's total rural families – keep dairy animals. Over half of the milk they produce, mainly buffalo milk, goes to feed people in the communities they live in, while one quarter of it is processed locally into yoghurts, ghee (clarified butter) and other dairy products.

India's dairy sector employs about 90 million people, of which 75 million are women. It is a significant source of income for small and marginal farmers, the landless poor and millions of rural families. And it is still India's biggest agricultural sector, contributing 22 per cent of total agricultural GDP. The country is the world's largest milk producer, accounting for more than 15 per cent of the total global dairy output. Milk is an essential part of Indians' diets. Almost all 108 million tonnes of dairy produced annually is consumed domestically.

Much is made of the significance of India's dairy co-operatives in the "white revolution", which saw a tripling of milk production between 1980 and 2006. But the real story lies with the people's milk sector, which still accounts for 85 per cent of the national milk market. It was India's small-scale farmers and domestic markets who were the real basis for the massive expansion in the country's dairy production over those years, and, as a result, the benefits of this boom in production have been widely distributed.

Through the 1980s and early 1990s, the National Dairy Development Board implemented the second and third phases of a program to increase milk production and consumption in the country. Operation Flood aimed to improve nutrition and reduce poverty by linking milk producers in Indian villages with urban markets. The program's success provided a steady income for farmers and even landless agriculture workers.

Table 1 Percentages of national milk markets handled
by the informal milk sector in selected countries

Country	Percentage of national milk market handled by the people's milk sector
All developing countries*	80
Argentina	15
Bangladesh	97
Brazil	40
Colombia	83
India	85
Kenya	86
Mexico	41
Pakistan	96
Paraguay	70
Rwanda	96
Sri Lanka	53
Uganda	70
Uruguay	60**
Zambia	78

* 85.4 per cent of the world's population lives in developing countries according
to the Human Development Index
**Figure is for cheese only
Source: GRAIN

Corporate hands off the people's milk

India's dairy sector has faced great challenges in recent years and the direction
taken so far in negotiations over free trade agreements, such as the India-EU
FTA, points to more difficulties ahead. With Western economies in crisis, India
represents potentially rich pickings for powerful transnational corporations.
The India-EU FTA essentially represents the demands of big corporations.

EUCOLAIT, the representative association of the European dairy trade and industry, is calling on EU negotiators to insist that the EU be given the same level of access to the Indian dairy market as India grants to other countries. In a December 2011 statement, EUCOLAIT highlighted India's position as the world's largest dairy consumer and said that India could become a consistent dairy importer. It says the EU should thus remain firm in negotiating an ambitious agreement delivering real market access for EU companies in the dairy sector.[1]

Against this, there is strong domestic support in India for the government to take adequate steps to protect the interests of the country's small dairy producers. On a parliamentary panel in April 2013, the members of parliament on the Committee of Agriculture stated that the interests of dairy producers in the country should be protected from monopolies and discriminatory and lopsided trade practices. There is also strong pressure from farmers' movements to halt the India-EU FTA, which will create further liberalization in the country and destroy India's agriculture sector, and the dairy sector in particular.

There is already a trend of increasing foreign investment in India's dairy, especially following the decision of the government in late 2012 to allow foreign retailers to own up to 51 per cent in multi-brand retail and 100 per cent in single brand retail. According to Kevin Bellamy from Rabobank, the world's largest agribusiness lender, this is the first step towards introducing outside dairy products into the Indian dairy market.

To get around the initial opposition to its foreign direct investment policies for retail, the Indian central government left the final decisions with the state governments. So far, of the country's 30 state governments, only 10 of them have stated they are fully in favor of the revised FDI policies, while seven states are opposed and the remaining ones have yet to take a position. The position of the state governments closely matches the strength of dairy co-operatives in the states. In some states, such as Karnataka, there is a high level of government involvement in the dairy co-operatives, which also provide an important source of government revenue.

The role of co-operatives in the Indian dairy sector

There are about 96,000 dairy co-operatives in India, ranging between the primary, district and state level. Karnataka is one of the states where the role of co-operatives has been crucial. The Karnataka Milk Federation (KMF) is the largest Co-operative Dairy Federation in South India, owned and managed by milk producers of Karnataka State. The KMF has more than 2.23 million milk producers in over 12,066 Dairy Co-operative Societies at village level, functioning under 13 District Co-operative Milk Unions in the state.

One of the district level co-operatives in Chamrajanagar receives 85,000 to 90,000 liters of milk a day from its 225 primary level co-operatives. The milk is collected twice a day from 60 bulk milk collection centers, on average one center for every five villages, and then transported to the chilling centers in Chamrajanagar town. There are three giant chilling tanks in the district with a capacity of 30,000 liters each.

Farmers who sell the milk to the co-operatives receive payment of between seven to 21 rupees per liter (11-34 cents), depending on the level of solid non fat (SNF) that is measured using lactometers available in each collection center. The average level of SNF in milk that the co-operatives receive is 8.4 per cent. The milk is then packed for sale, under the Nandini brand. The Karnataka Milk Federation sells the milk to consumers for 32 rupees per liter.

Karnataka produces a total of five million liters of milk annually, easily exceeding the state's consumption of three million liters. The surplus is processed into powdered milk (skim milk powder and whole milk powder) by SKA Dairy Foods, a private company contracted by the government through a tender process. The co-operative in Chamrajanagar has huge storage with capacity of 85,000 kilograms of milk powder. As the selling price of milk powder is 10 times the price of fresh milk, milk powder is an important product of the co-operative, along with ghee. Part of the milk powder production is used by the government of Karnataka for its school-children food subsidy program, which provides skim milk powder for children in grades one through six and whole milk powder to class seven to ten. The remaining surplus is then sold to other states such as Tamil Nadu, and even to Delhi. It is a major source of income for the state.

With its extensive coverage of villages in Karnataka, the KMF is able to sustain income for small milk producers throughout the state. The co-operative system is not flawless. One of the most common problems is late payment to farmers. Producers are supposed to be paid each week, but one farmer near Rajarajeswar Nagar village said payments can be held up for more than a month. Financial transparency between various levels of the co-operatives and among members is also an issue. Only one representative from the lower levels of a co-operative participates at the level above.

The Karnataka co-operative and others have huge capacity to collect milk directly from farmers and bring it to market. Mega dairy farms, in contrast, have little interest in doing this. Instead, many big dairy companies seek to import powdered milk from Europe or New Zealand to meet demand.

The co-operatives have also played a crucial role in resisting the India-EU FTA. Amul, a major dairy co-operative from the state of Gujarat, wrote letters to the Minister of Commerce expressing its strong opposition to granting any kind of advantage in terms of import duty on dairy products.

Dairy and India's food sovereignty

"We don't really intend to be dairy farmers, but it is part of our life," Marayal says. For farmers like her, cows and buffaloes provide a steady and sustainable income.

The vibrant network of small producers and milk co-operatives that makes up most of India's dairy sector is a powerful model: one which is now threatened by free trade agreements and liberalized investment policies.

Opening up access to import heavily subsidized milk powder and other milk products from the European Union will allow processors and retailers to put downward pressure on local milk prices, forcing farmers to accept prices below the costs of production.

This is why India's farmers, co-operatives and trade unions have been at the forefront of protesting against the EU-India free trade agreement over the past year. They understand that high tariffs are a necessity. Far from leading to higher prices for consumers, such tariffs will protect against dumping and prevent big processors from substituting cheap, processed dairy – or even non-dairy – products for real milk.

Investors and big dairy corporations are working hard to hijack dairy markets in India and across the South. In addition to its interests in India, Cargill is investing hundreds of millions in mega dairy farms in China. Fonterra is also expanding aggressively in China and Brazil. If they succeed, it would spell economic and social disaster for millions of people.

But experiences elsewhere show that people's milk can successfully resist the powerful forces lined up against it. In Colombia, small producers, vendors and consumers formed an alliance that forced the government to recognize people's milk – *leche popular* – as legal and essential. This success was built on three key arguments. First, that people's milk presently meets the bulk of dairy needs and Big Dairy cannot replace it. Second, that millions of people's livelihoods depend on small dairy production; here too, Big Dairy offers no alternative. And finally, that the system of people's milk provides safe, fresh, nutritious milk at affordable prices to millions of households.

This is the system that needs to be defended as a cornerstone of food sovereignty in India, Colombia and elsewhere. Milk must remain in the hands of the people.

Box 1: Make way for mega dairy farms

India's new FDI and trade policies not only open the country up to dairy imports, they also facilitate the takeover of local dairy production and processing. In 2011, the Carlyle Group, one of the largest private equity firms in the US, bought a 20% stake in Tirumala Milk Products, a private dairy company that handles 1.2 million liters of milk daily from its procurement and distribution network in Andhra Pradesh, Karnataka and Tamil Nadu. A year later, the French dairy giant Danone began negotiating the purchase of a controlling stake in Tirumala. That same year, Rabobank made an $18.5 million equity investment in Prabhat Dairy of Maharashtra through its India Agribusiness Fund. In August 2013, Rabobank invested another $12 million in Prabhat while the French development finance institution Proparco put in $9 million.

Part of this foreign investment in dairy production is driven by broader foreign investment in the food sector. As companies like McDonalds enter India, so do their main global suppliers. When McDonalds began opening restaurants in India in the late 1990s, its main dairy supplier, US-based Schreiber Foods, created a partnership with the wealthy Goenka family to establish a large dairy-processing company in Maharashtra, now called Schreiber-Dynamix.

The company set up contract farming and collection centers to collect milk from local farmers, but also began building up its own large-scale farm to supply its needs. In November 2010, the company inaugurated a "future ready" 6,000-cow dairy farm on 120 ha in Pune District, with backing from the State Bank of India. Dynamix also supplies Danone, Nestlé, Yum! and Kentucky Fried Chicken. In February 2013, Nestlé made direct investment

---

in a milk collection company linked to Schreiber-Dynamix by acquiring a 26-per-cent stake in Indocon Agro and Allied Activities Pvt Ltd, which is engaged in the milk collection business in western India.

Among the private dairy companies, the trend is clearly towards the creation of vertically integrated supply chains, starting with their own mega farms. The world's biggest milk producer, Fonterra, has a joint venture with the Indian Farmers' Fertiliser Co-operative (IFFCO) and the Indian financial company Global Dairy Health (GDH) to establish a 13,000 cow dairy farm on 65 ha of land in an IFFCO Special Economic Zone near Nellore, Andhra Pradesh. The project now appears to be on hold after the Andhra Pradesh Animal Husbandry Department rejected the companies' application to import 9,000 high yielding pregnant cows from New Zealand. But a master plan for the special economic zone was approved in 2012 and Kalyan Chakravarthy of GDH was appointed as its executive director.

The GDH also has plans for mega dairy farms in other locations. In a December 2010 presentation it described three farm projects that it was pursuing: the one in Nellore with IFFCO and Fonterra, a second for 3,500 cows in Bangalore with a "strategic local partner" with a target for operations starting in 2011, and a third for 3,500 cows with a local partner in North Coastal Andhra Pradesh, on the border with Orissa.

The transnational grain trading corporation, Cargill, also has plans to enter the Indian dairy sector. In 2010, Cargill announced that it would be investing in dairy farms in China and India through its hedge fund Black River Asset Management. Later in 2012, Black River's subsidiary, Cargill Ventures, made its first investment in the Indian dairy sector. It invested $15 million in Dodla Dairy, also based in Andhra Pradesh, in Nellore. Dodla initially had investment from Indian private equity firm Ventureast.

Box 2: Sri Lanka: Natural milk production – A people's concern, not of big business

People's participation in Sri Lankan milk production is still around 53 per cent.

If the government of Sri Lanka is consistent with its policy of National Milk Production, it should end the importation of powdered milk. Immediate steps have to be adopted to concentrate on local production of fresh milk mainly for children and the sick. Even rationing of milk is advisable till we reach a glut of production.

The import of foreign cows and genetic resources must also be stopped. The available cattle population (both cow and buffalo) – properly fed and protected – will be sufficient to initiate a new beginning. Even the Bos Indicus breed from India should not be imported. There are enough hybrid animals – Saheewal and Gir, for milk, and Khillari as draught animals – in Sri Lanka.

A Sri Lankan Bos Indicus cow can give at least five bottles of milk (1 bottle = 750 ml) or more per day. This animal thrives on various grasses and herbs. No concentrate feed is necessary at all. Therefore, it is a low cost production of milk full of cream and proteins. Buffalo populations of both the local and Indian breeds should be developed. The Sri Lankan Bos Indicus breed should be utilized mainly for integrated agriculture and for plowing and rural transport. Cow dung and cow urine are the most essential food for the soil – and the soil biota. The Indian farmers call it Jeewa Amurthaya – the source of wealth – and the power of a nation will depend on its healthy soil, its healthy plants and healthy living beings.

People's Milk is an engine of poverty alleviation and health. It provides livelihoods and safe, affordable, nutritious foods. The revenues earned are distributed evenly and consistently throughout the sector. Everyone wins with people's milk, except for big business. And this is why there is such pressure to destroy it. What does Big Dairy have to offer? Instead of fresh, high-quality milk produced and supplied in the most sustainable ways, we are offered powdered and processed milk produced on highly polluting mega farms and sold in all kinds of packaging, at double the cost!

Linus Jayatilake, National Movement of Sri Lankan Dairy Farmers

Full article: https://www.grain.org/e/4873

1 EUCOLAIT and EDA (2013) "EU-India FTA – Position of Dairy Sector", 12 August

4.3 Food sovereignty for sale: supermarkets and dwindling people's power over food and farming in Asia

In the past decade or so, corporations have been taking over a bigger and bigger slice of the production, distribution and sale of food across Asia. This is having a major impact on the region's small-scale traders and processors, its fresh markets and street vendors. Corporate supermarkets are expanding faster in Asia than anywhere else on the planet. And as supermarkets and their procurement chains expand, they take revenue out of traditional food systems – and out of the hands of peasants, small-scale food producers and traders. They also exert increasing influence over what people eat and how that food is produced.

Asia continues to rely on traditional food systems for most of its food supply. But the entry and aggressive expansion of transnational food corporations, beverage companies and supermarket chains over the past decade has had major impacts on Asia's farmers, food workers, traders and consumers.

Opening the flood gates to supermarket expansion

More food is consumed in Asia than anywhere else in the world. So it is no surprise that the continent has become a major focus for transnational food retailers looking to expand their businesses. Asia is now the fastest-growing market in the world for corporate food retail, and the industry's preferred investment destination.

The expansion of supermarkets in Asia is being driven by the same factors as in other regions: income growth and rapid urbanization on the demand side and marketing and foreign direct investment (FDI) on the supply side. Retailers are using different store formats (from wholesale outlets to hypermarkets to convenience stores) to avoid restrictions on foreign investment or municipal zoning laws and to maximize their reach.

This supermarket expansion, however, could not have happened without the wave of investment liberalization that preceded it. India began opening its retail sector to foreign investment in 2006 by allowing 100-per-cent FDI in cash and carry wholesale trading. Then in December 2013, despite tremendous public protest, India's central government enacted new FDI policies allowing foreign retailers to own up to 51 per cent of multi-brand retailers and 100 per cent of single brand retailers, which, like IKEA or Apple, only sell their own product brands. Implementation, however, was left to each state government. Today, fresh markets still account for 98 per cent of total food retail sales but the "organized" or "modern" food retail sector is growing rapidly. The number of modern retail outlets increased from an estimated 200 outlets in 2005 to 3,000 outlets in 2012.

Multilateral and bilateral free trade and investment agreements have also opened the door to FDI in the retail sector. In China, for example, there were no supermarket chains in 1989. But with a progressive liberalization of foreign direct investment in retail that started in 1992 and culminated in 2004 – as a provision of accession to the World Trade Organization – China's supermarket sector has grown between 30 and 40 per cent per year, the fastest rate of growth in the world.[1]

Although foreign investment plays a crucial role in the growth of supermarkets in Asia, state or domestic investment is also significant. In Indonesia, for example, the number of retail outlets in Indonesia grew, on average, by almost 2,000 new outlets per year, from 10,365 outlets to 18,152 outlets between 2007 and 2011. The number of hypermarket outlets jumped from 99 to 167 over the same period. Nearly all of these stores are controlled by just a handful of powerful retail groups. Trans Retailindo, Hong Kong-based Dairy Farms, 7 Eleven, Gelael Group, and the two most aggressive locally owned retail chains, Indomart and Alfamart. Other than Dairy Farm, all of these groups are state-owned or joint ventures that receive financial support from foreign companies.

In India too, the corporate retail sector remains dominated by large Indian companies. Their growth has been enabled by national and local regulations and development programs that seek to replace fresh markets with supposedly safer and more hygienic corporate retailers.

Supermarketization, changing the face of the Asian market

Across the region, fresh markets provide consumers with fresh quality vegetables, fruits, meats and other food. These markets also provide livelihoods to millions of people along the distribution chain, from the small farmers who bring in their harvests, to stall owners, to the street vendors and a vast range of other informal workers including porters and loaders.

In Indonesia alone, there are 12.5 million stall owners operating in the 13,450 fresh markets registered in the country, and this figure does not include the many informal workers earning income from these markets.

Supermarkets pose a direct threat to the livelihoods of these people. As supermarkets expand, they capture an increasing share of the national expenditures on food, leaving the millions of people who depend on fresh markets and small retail shops with less to share amongst themselves. A direct result of this in Indonesia, for example, is that the number of fresh markets in the country is shrinking by an average of 8.1 per cent every year. The Indonesia Market Traders Union (IKAPPI) says that more than 3,000 fresh markets, each with dozens of kiosks, were shut down between 2007 and 2011, with the overall number of fresh markets declining from 13,450 to 9,950.[2] When asked, nearly half of the traders said direct competition with supermarkets was their reason for having to close down their stalls.[3]

This explains why street vendors and informal traders have been at the forefront of resistance to the liberalization of FDI in retail in Asia. In India, where almost 40 million people still rely on the informal trade sector and fresh markets, the resistance is fierce. In 2006, a national steering committee was created to co-ordinate a movement for "retail democracy" called Vyapar Rozgar Bachao Andolan, led by those who have been most affected by retail liberalization: trade associations, unions, hawkers' organizations, farmers' groups and small-scale industries. On 5 February 2014, thousands of street vendors marched on the Indian Parliament pushing for the adoption of a Street Vendors Bill and the reversal of national policies that allow foreign companies to invest in the retail sector.[4]

No place for small farmers on supermarket shelves

Asia's small traders sit at the front end of the local food systems that ensure the procurement and distribution of food grown on millions of small farms across the region. These traders typically procure their fresh fruit and vegetables, meat, eggs and fish from wholesale markets to which nearby farmers bring their produce daily.

Corporate retailers rely on totally different systems of procurement and distribution. Each supermarket chain co-ordinates its own procurement of products centrally for all of its outlets around the world. Foods are supplied by large transnational companies that can consistently supply large volumes according to exacting standards set by supermarkets.

There is very little room for small farmers to participate in these integrated supermarket supply chains. One of the main problems is that supermarkets demand adherence to standards for food safety that are impossible for small farmers to comply with.

Over the past decade, the global food industry has developed hundreds of schemes to regulate the safety and quality of products that move through their systems. In 1999, a group consisting of 17 European retailers decided to create their own verification system for suppliers and developed standards – covering the production of fruits, vegetables, grains like wheat, barley and canola, livestock, animal feed, flowers and more – that came to be known as the Good Agricultural Practices (GAP).[5]

The GAP standards initiated by retailers have since been promoted through the United Nations' Food and Agriculture Organization (FAO) and the Codex Alimentarius for translation into national regulations. The standards are officially voluntary, but governments and big food retail chains are increasingly making GAP standards compulsory, not only for the sale of products to retailers, but also for farmers wanting to access extension, marketing and credit programs. Official documents from the FAO and governments indicate that the goal is to make these standards legally binding.[6]

Governments in Asia are making special efforts to ensure that small farmers follow GAP standards. But the standards are not at all adapted to the farming systems of most small farmers. GAP standards typically include requirements such as storage rooms with solid walls and cemented floors; potable water for handling products after harvesting (and, in some cases, even for irrigation); strict record keeping of all activities, sales and purchases, use of commercial seeds and other inputs; and hired technical assistance by agronomist or other professionals. GAP standards ban animals from crop fields, and spell out requirements for personal hygiene.

These standards were developed in Europe, and have no relationship to the traditional food systems of Asia. Carrefour's Indonesia fresh products merchandise director told GRAIN that its suppliers must comply with the company's internal procurement standards book. Although Carrefour is now wholly owned by an Indonesian company, Trans Retailindo, the standards book is unchanged. It is almost impossible for small-scale Indonesian farmers to comply with these European standards, without access to the farm machinery and advanced post-harvest technology. The standards also include precise norms for freshness and product sizes that are suited to industrial agriculture; for example, broccoli must have a bright green color with exactly five centimeter of stem.

Across Asia, compliance for most small farmers is simply impossible or far too costly. The "solution" often proposed by governments and the food industry is more vertical integration, especially contract farming, so that farmers can concentrate on following GAP in the field and the companies they supply can take over all forms of handling, processing and marketing. This, of course, comes at a cost for farmers, both in terms of higher expenses – since food companies charge for every service – and in terms of loss of control over the marketing process, frequently resulting in low and seriously delayed payments for produce.

In India, many national and transnational companies have moved to set up contract farming arrangements to supply supermarkets. In Bangalore, for instance, farmers are being drawn into contract arrangements with companies supplying supermarkets with promises of guaranteed markets, stable and higher prices and technical assistance. But farmers there say the terms and conditions they must follow are too complicated and onerous. They also say that their harvests are often rejected and go to waste, and that payments by the contractors are regularly late. They have a hard time dealing with the pace and scale of production based on the contract terms, and are shut out of all production decisions, for example over the type of crops and inputs, as well as decisions regarding sales, such as gauging the quality of their crops and the price they are due.[7]

"I have been growing vegetables and selling it directly to the consumers in a nearby market. I do not know how to sell it to Metro," says Rudresh, a farmer from Hoskote, a rural district in Bangalore. "They only buy the top quality produce, but in the local market I sell all my vegetables, at varying prices, according to the quality."[8]

The reality is that even with the increasing number of contract farming programs, supermarkets source very little of their production from small farmers. Most of their food supplies come directly from large corporate farms. The Thai company Charoen Pokphand is one of the leading suppliers of meat to global supermarkets. Its operations are vertically integrated, from breeding farms to slaughterhouses to food processing plants. The company even operates its own chain of supermarkets (CP Fresh Mart) and convenience stores (7 Eleven).[9]

Wal-Mart's largest global meat supplier is US-based Tyson Foods, the world's largest meat producer. Tyson is currently spending hundreds of millions of dollars to set up its own operations in China. The company, which very recently had no farms in China at all, plans to construct 90 chicken farms there by 2015.

Food safety for sale

In 2011, GRAIN released a report, "Food safety for whom? Corporate wealth vs people's health", which shows how trade agreements have become the core mechanism to expand and enforce food safety standards around the world. As agriculture markets have been profoundly liberalized, there has been a boom in the global trade in food.

Too often, the food safety rules that emerge from trade negotiations become mechanisms to force open markets, or backdoor ways to limit market access; they do little to protect public health, serving only corporate growth imperatives and profit margins. There is no evidence that standards like GAP actually improve the quality of food or reduce the possibility of outbreaks of food-borne diseases.

In fact, global supply chains make consumers more susceptible to food contamination. A small farm that produces some bad meat will have a relatively small impact. A global system built around geographically concentrated factory-sized farms does the opposite: it accumulates and magnifies risk, subjecting particular areas to industrial-style pollution and consumers globally to poisoned products.

Food safety and standards are partly a response to consumer demands, but they are also aggressively promoted as premium products by corporate retailers. Green certification and eco-labelling programs, for instance, represent a market response to the demand for environment-friendly practices and healthy products. Eco-labelling attempts to capitalize on the price premiums consumers are willing to pay for both the private good of safe food and the public good of an improved environment.

Food sovereignty at stake

Mega retailers want to offer the same fresh fruits and vegetables all year round, whether they are in season or not. They are able to do so by sourcing produce from different geographic locations around the world. But they also want products as cheaply as possible. So they look for production centers where they can source at the lowest cost. China, for instance, is becoming a major production and distribution center for poultry and horticulture products for supermarkets in many countries in Asia.

The growing number of free trade and investment agreements in Asia facilitates global procurement systems for retailers. Since coming into effect in January 2010, the ASEAN-China Free Trade Agreement (ACFTA), one of the most controversial trade agreements in the region, allows zero tariffs for more than 600 agriculture products from China to Southeast Asian countries.

Supermarkets will also reap big benefits from the Asean Economic Community (AEC), which came into being in 2015. The AEC fully integrates Southeast Asian countries according to five core elements: free flow of goods, free flow of services, free flow of investment, free flow of capital, and free flow of skilled labor. The Thailand-based wholesale chain Siam Makro has already set up new outlets along the Cambodian border not just to target new consumers but also to benefit from neighboring country suppliers that might offer lower prices than Thai producers.[10] Siam Makro chief executive Suchada Ithijarukul confirms that they are also seeking these opportunities in Laos.[11]

Global procurement allows retail chains to undercut local production by sourcing from cheap production centers for the lowest prices. This puts pressure on local producers and increases monoculture production in specific areas. There is no better example of this process than vegetable oils. In the last two decades, world oil crop production has been dominated by only three oils – soybean, oil palm and canola/rapeseed. In 2012-13 world oil crop production of these three oils accounted for 76.7 per cent of total production.[12]

The growth of these vegetable oils is highly concentrated in specific areas of low-cost production – Brazil and Argentina for soybeans, Malaysia and Indonesia for oil palm and Canada and China for canola. The dominance of these vegetable oils has undermined the viability of other oil crops such as coconut, groundnut, sunflower, cottonseed and olive. Import tariffs on vegetable oils in India were reduced three times under pressure from World Bank Structural Adjustment Programme, falling from 65 per cent in 1994 to 20 per cent in 1996 and then to 15 per cent in 1998.[13]

The effect on edible oil producers in India has been catastrophic. Over 10 years, the price paid to the four million coconut farmers in the state of Kerala, for example, dropped from 10 rupees a piece to just two to three rupees a piece. The fatal blow came in April 2008 when the import duty on all crude edible oils was reduced to zero.[14]

The suicide rate among farmers soared. According to official data, nearly 160,000 Indian farmers have committed suicide over the last decade: one every half an hour. Debt linked to failed GM cotton crops and the free-falling price of vegetable oils are the two leading causes.[15]

Nutrition transition and changing dietary patterns

One of the most important challenges that Asia faces today is how to feed its large and growing urban populations. Urbanization brings with it changes in life styles and consumption patterns, marked by rising demand for semi-processed or ready-to-eat foods. Supermarket chains are positioning themselves to take advantage of this situation and become the main suppliers of food to the region's urban centers.[16]

Vertically integrated food supply chains linking producers, processors, distributors and retailers become essential for meeting the changing demand.

Food consumption patterns are shifting toward more meats or fats, dairy products, and sugary foods, as the effects of globalization and the international food trade reshape the types of food produced, and increase the amount of food imported into developing countries. The traditional diets of many communities in Asia, rich in starch-based calories, are shifting towards more Western diets, high in sugars, fat, and animal-source food.[17] Cheap packaged and processed food is replacing healthier daily meals of fresh food for poorer households. Supermarkets stimulate the shift with advertising and promotions (see Chapter 4.6 on free trade and Mexico's junk food epidemic), even as they establish organic and healthy product niches at higher prices. A survey published in the *Journal of the American Dietetic Association* compared the prices of 370 foods sold at supermarkets, and found that processed or "junk" food costs as little as $1.76 per 1,000 calories, whereas fresh fruits and vegetables cost more than 10 times as much for the same number of calories.[18]

Processed foods typically have little nutritional value. High levels of saturated trans-fatty acids, salt and sugar that help with preservation and enhance flavors are linked to obesity and diet-related diseases such as diabetes, high cholesterol and heart disease. Thus, as supermarkets take over more of the food supply, consumption of these processed foods goes up, as does obesity and other food-related health problems.[19]

An analysis of public data about what the world eats found that between 1980 and 2008, the number of overweight and obese people in the developing world more than tripled, from 250 million to 904 million. In the same period, the number of overweight and obese people in developed countries increased "only" 1.7 times.[20] Half of the world"s obese people live in just 10 countries, and China, India, Pakistan and Indonesia – the Asian countries with fastest supermarket expansion – all feature on that list.[20]

More than half of China's 1.3 billion people now live in urban areas, up from around 400 million just a decade ago. Supermarket chains are rapidly expanding in the country's major cities, playing the role of food suppliers and transforming dietary patterns.

China's urban population mostly eats at home, so supermarkets like Carrefour, Wal-Mart and their Chinese-owned counterparts are the main drivers of changes in dietary patterns, rather than Western-style fast-food chains.[21] Supermarkets are growing faster in China than anywhere else in the world, and this expansion is now spreading into smaller cities and towns, and even reaching higher-income earners in rural areas.

This growth coincides with a dramatic shift away from the consumption of grains and complex carbohydrates and towards the consumption of meats and fats. Increased consumption of processed foods also means increased consumption of fats, and particularly palm oil, the world's discount source of fats. It is estimated that palm oil is found in half of all packaged foods on supermarket shelves.[22] In China, the annual per-capita consumption of vegetable oils rise from 3 kilograms in 1980 to 23 kilograms in 2009 – that is roughly 64g per day, almost twice the fat intake required to meet a person's nutritional requirements. Palm oil now accounts for a third of the vegetable oil consumed in China, nearly three times the share it held in 1996.[23]

The impact of this dietary change is quickly becoming visible, with the number of obese people in China rising from 18 million to almost 100 million between 2005 and 2011. Between 1992 and 2002, the percentage of overweight people in China rose by nearly 40 per cent (its obesity rate nearly doubled) according to the Chinese Centre for Disease Control and Prevention – coinciding neatly with the liberalization of FDI in retail over the same period.[24]

Feeding growing populations?

The current global food distribution system is unsustainable and undermines food sovereignty. The expansion of supermarkets puts small farmers in direct competition with industrial agriculture, and also has negative impacts on local markets and communities. As more and more people in Asia turn to supermarkets for their food, food diversity is eroded and corporate supermarkets gain more power to determine food systems, from production up to distribution chains and consumption.

The shift towards supermarkets cannot be seen as a solution to feeding growing populations in Asia. It is transferring control over and access to food from millions of small farmers, home food artisans, local food markets, and consumers to a handful of corporations like CP, Aeon, Dairy Farm, Wal-Mart, and other global retailers, and their corporate suppliers from the food industry and agribusiness. It puts at risk the livelihoods of hundreds of millions of people who rely on the food sector for their livelihoods.

Across the region, there is both growing awareness of the threat posed by global retailers and a growing resistance against their expansion. But we must continue envisioning and building strategies and alternatives to the supermarket model of food distribution, in order to move forward in a way that strengthens social, community-based and public food systems and assures the survival of small food producers and local markets.

Full article available at: https://www.grain.org/e/5010

1 Reardon, T. and Berdegué, J. (2008) "The Retail-Led Transformation of Agrifood Systems and its Implications for Development Policies", World Bank, https://openknowledge.worldbank.org/handle/10986/9233

2 Berita Moneter (2014) "Pasar tradisional menggugat", http://www.beritamoneter.com/pasar-tradisional-menggugat/

3 SMERU Institute (2007), "Dampak supermarket terhadap pasar dan pedagang ritel tradisional di daerah perkotaan di Indonesia", pdf, http://www.smeru.or.id/sites/default/files/publication/supermarket_ind.pdf

4 Dharmendra, K. and Ranjan, V. (2011) "Corporatizing Agri- Retail: implications for weaker links of the food supply chain. An appraisal of Bengaluru", India FDI Watch, pdf, http://indiafdiwatch.org/wp/content/uploads/2013/03/India_FDI_Watch_Booklet_in_English1.pdf

5 See for example: http://www.fao.org/prods/gap/

6 Future Unidroit/FAO (2014) "Legal guide on contract farming", http://www.unidroit.org/english/guides/2015contractfarming/brochure-e.pdf, p.10-11. The document clearly states

that agricultural producers are bound to apply good practices based on three categories of obligations relating to receiving, taking care of and using inputs according to the guidelines given by the contractor.

7   Dharmendra, K. and Ranjan, V. (2011), *op cit.*

8   Metro AG, known as Metro Group, is a German global diversified retail and wholesale/cash and carry group. Established in 1964, it is now the fifth-largest retailer in the world measured by revenues.

9   See: Charoen Pokphand Foods PCL, http://www.cpfworldwide.com/en/about/

10   Interview with Siam Makro procurement manager, March 2014.

11   Jitpleecheep, P. (2014) "Makro plans Asean drive: big retail chain tests market in Myanmar," *Bangkok Post*, 21 May, http://www.bangkokpost.com/business/retail/410860/makro-plans-asean-drive

12   USDA (2013), Oil Crops Outlook, March

13   Dohlman, E. et al. (2003) "India's edible oil sector: Imports fill rising demand," USDA, http://www.ers.usda.gov/media/1724067/ocs090301.pdf

14   Jafri, A. (2011) "Trade Liberalisation's Impact on Edible Oil Sector in India", Focus on the Global South, http://siccfm.blogspot.ca/2012/01/trade-liberalization-and-impact-on.html

15   Keck, Z. (2013) "Why Do So Many Indian Farmers Commit Suicide?" *The Diplomat*, 19 July, http://thediplomat.com/2013/07/why-do-so-many-indian-farmers-commit-suicide/

16   FAO (2009, "How to feed the world in 2050", 12 October, http://www.fao.org/fileadmin/templates/wsfs/docs/expert_paper/How_to_Feed_the_World_in_2050.pdf

17   Hawkes, C. (2007) *Globalization and the Nutrition Transition*, Cornell University

18   Parker-Pope, T. (2007) "A high price for healthy food", *New York Times*, 5 December, http://well.blogs.nytimes.com/2007/12/05/a-high-price-for-healthy-food/?_php=true&_type=blogs&_r=2

19   Popkin, B.M. (2006) "Global nutrition dynamics: the world is shifting rapidly toward a diet linked with noncommunicable diseases", *American Journal of Clinical Nutrition*, 84(2):289–298

20   Viegas, J. (2014) "Top 10 countries with the most obese people named," *Discovery News*, 28 May, news.discovery.com/human/genetics/top-10-countries-with-the-most-obese-people-named-140528.htm

21   Popkin, B.M. (2008) "Will China's Nutrition Transition Overwhelm Its Health Care System And Slow Economic Growth?" *Health Affairs*, http://content.healthaffairs.org/content/27/4/1064.abstract

22   Rainforest Action Network (2013), "Conflict palm oil", September, http://www.ran.org/palm_oil

23   Sharma, D.C. (2012), "Rise in oil consumption by Indians sets off alarm", *India Today*, 2 April, http://www.ran.org/palm_oil

24   Patterson, S. (2011) "Obesity in China: Waistlines are Expanding Twice as Fast as GDP." *US-China Today*, 4 August, http://www.uschina.usc.edu/w_usct/showarticle.aspx?articleID=16595&AspxAutoDetectCookieSupport=1

4.4 How does the Gates Foundation spend its money to feed the world?

Since the Bill and Melinda Gates Foundation added "feeding the world" to its objectives almost a decade ago, it has channelled an impressive three billion dollars towards agricultural projects, much of it to improve farming in Africa. From nowhere on the agricultural scene less than a decade ago, the Gates Foundation has emerged as one of the world's major donors to agricultural research and development. But GRAIN analyzed the foundation's agricultural grants records for the past decade and came to some sobering conclusions.

The foundation may say it's fighting hunger in the South, but its money is overwhelmingly sent to the North. The bulk of its funding goes to high-tech scientific outfits rather than to supporting the solutions that farmers themselves are developing on the ground. The Gates Foundation also uses its money to push for legislation and policies to open up markets to foreign corporations, to privatize land and seeds and to allow for the introduction of GMOs.

"Listening to farmers" is a stated guiding principle for the Gates Foundation, yet when we follow the money, Africa's farmers are rather cast as recipients, mere consumers of knowledge and technology from others.

Here are some of the conclusions we were able to draw from the data.

The Gates Foundation fights hunger in the South by giving money to the North

Graph 1 on page 187 gives the overall picture. Roughly half of the foundation's grants for agriculture went to four big groupings: the Consultative Group for International Agricultural Research (CGIAR) – a global agriculture research network; international organizations (World Bank, UN agencies, etc.); Alliance for a Green Revolution in Africa (AGRA – set up by Gates itself) and the African Agricultural Technology Foundation (AATF). The other half ended up with hundreds of different research, development and policy organizations across the world. Of this last group, more than 80 per cent of the grants were given to organizations in the US and Europe, 10 per cent went to groups in Africa, and the remainder elsewhere. By far the main recipient country is Gates's home country, the US, followed by the UK, Germany and the Netherlands.

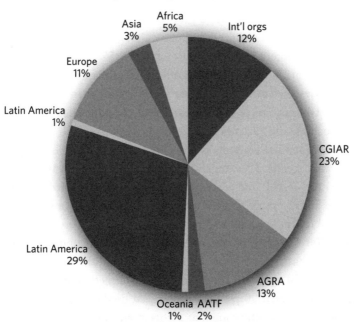

Graph 1 The Gates Foundation's $3 billion pie
(agriculture grants, by region)

When it comes to agricultural grants by the foundation to universities and national research centers across the world, 79 per cent went to grantees in the US and Europe, and a meagre 12 per cent to recipients in Africa.

The North-South divide is most shocking, however, when we look at the NGOs that the Gates Foundation supports. One would assume that a significant portion of the frontline work that the foundation funds in Africa would be carried out by organizations based there. But of all the money that the Gates Foundation has granted to non-governmental organizations for agricultural work, more than three-quarters has gone to organizations based in the US. Africa-based NGOs get a meagre four per cent of the overall agriculture-related grants to NGOs.

The Gates Foundation gives to scientists, not farmers

The single biggest recipient of grants from the Gates Foundation is the CGIAR, a consortium of 15 international agricultural research centers. In the 1960s and 1970s, these centers were responsible for the development and spread of a controversial green revolution model of agriculture in parts of Asia and Latin America – a model focused on the mass distribution of a few varieties of seeds that could produce high yields – with the generous application of chemical fertilizers and pesticides.

The CGIAR centers have received more than $720 million from Gates since 2003. During the same period, another $678 million went to universities and national research centers across the world – more than three-quarters of them in the US and Europe – for research and development of specific technologies, such as crop varieties and breeding techniques.

The Gates Foundation's support for AGRA and the AATF is tightly linked to this research agenda. These organizations seek, in different ways, to facilitate research by the CGIAR and other research programs supported by the Gates Foundation and to ensure that the technologies that come out of the labs get into farmers' fields. We could find no evidence of any support from the Gates Foundation for programs of research or technology development carried out by farmers or based on farmers' knowledge, despite the multitude of such initiatives that exist across the continent. (African farmers, after all, do continue to supply an estimated 90 per cent of the seed used on the continent!) The foundation has consistently chosen to put its money into top-down structures of knowledge generation and flow, where farmers are mere recipients of the technologies developed in labs and sold to them by companies.

The Gates Foundation buys political influence

Does the Gates Foundation use its money to tell African governments what to do? Not directly. The Gates Foundation set up the AGRA and describes it as the "African face and voice for our work". The AGRA, like the Gates Foundation, provides grants to research programs. It also funds initiatives and agribusiness companies operating in Africa to develop private markets for seeds and fertilizers through support to "agro-dealers" (see Box 1, Gates and AGRA in Malawi, page 189). An important component of its work, however, is shaping policy.

AGRA intervenes directly in the formulation and revision of agricultural policies and regulations in Africa on such issues as land and seeds. It does so through national "policy action nodes" of experts, selected by AGRA, that work to advance particular policy changes. For example, in Ghana, AGRA's Seed Policy Action Node drafted revisions to the country's national seed policy and submitted it to the government. The Ghana Food Sovereignty Network has been fiercely battling such policies since the government put them forward. In Mozambique, AGRA's Seed Policy Action Node drafted plant variety protection regulations in 2013, and in Tanzania it reviewed national seed policies and presented a study on the demand for certified seeds.

The AATF is another Gates Foundation-supported organization that straddles the technology and policy arenas. It has received $95 million from the Gates Foundation, which it used to develop and distribute hybrid maize and rice varieties, but also to "positively change public perceptions" about GMOs and to lobby for regulatory changes that will increase the adoption of GM products in Africa. In a similar vein, the Gates Foundation has provided grants to Harvard, Michigan and Cornell universities in the US to help African policy-makers decide on how best to use and promote the new biotechnologies.

Listening to farmers?

"Listening to farmers and addressing their specific needs" is the first guiding principle of the Gates Foundation's work on agriculture. But it is hard to listen to someone when you cannot hear them. Small farmers in Africa do not participate in the spaces where the agendas are set for the agricultural research institutions, NGOs or initiatives that the Gates Foundation supports. These spaces are dominated by foundation representatives, high-level politicians, business executives, and scientists.

Listening to someone, if it has any real significance, should also include the intent to learn. But nowhere in the programs funded by the Gates Foundation is there any indication that it believes that Africa's small farmers have anything to teach, that they have anything to contribute to research, development and policy agendas. The continent's farmers are always cast as the recipients, the consumers of knowledge and technology from others. In practice, the foundation's first guiding principle appears to be a marketing exercise to sell its technologies to farmers. In that, it looks, not surprisingly, a lot like Microsoft.

> **Box 1: Gates & AGRA in Malawi:**
> **Organizing the agro-dealers**
>
> One of the core programs of the Gates supported Alliance for a Green

Revolution in Africa (AGRA) is the establishment of agro-dealer networks: small, private stockists who sell chemicals and seeds to farmers. In Malawi, AGRA provided a $4.3 million grant for the Malawi Agro-dealer Strengthening Programme (MASP) to supply hybrid maize seeds and chemical pesticides, herbicides and fertilizers.

The main supplier to the agro-dealers in Malawi has been Monsanto, responsible for 67 per cent of all inputs. A Monsanto country manager disclosed that all of Monsanto's sales of seeds and herbicides in Malawi are made through AGRA's agro-dealer network.

"Agro-dealers... act as vessels for promoting input suppliers' products," says one MASP project document. Training the agro-dealers on product knowledge is carried out by the corporate suppliers of the products themselves. In addition, these agro-dealers are increasingly the source of farming advice to small farmers, and an alternative to the government's agricultural extension service. According to the World Bank: "The agro-dealers have become the most important extension nodes for the rural poor. A new form of private sector driven extension system is emerging in these countries."

The agro-dealer project in Malawi has been implemented by CNFA, a US-based organization funded by the Gates Foundation, USAID and DFID. Its local affiliate is the Rural Market Development Trust (RUMARK), whose trustees include four seed and chemical suppliers: Monsanto, SeedCo, Farmers World and Farmers Association.

Adapted from "The hunger games" by War on Want, London, 2012.

GRAIN's fully referenced report and database, on which this article is based, is available from its website: http://www.grain.org/e/5064

4.5 Planet palm oil: peasants pay the price for cheap vegetable oil

Palm oil is ubiquitous in our food systems. Look at the ingredients on any packaged food, and chances are you will find it there. Food companies love it, because it's cheap and abundant, so they use it whenever they can.

Demand is set to grow even further, as free trade agreements come on line that make it easier to import palm oil as a substitute for local animal or vegetable oils, as transnational food companies and supermarkets expand sales of processed and packaged foods in the South, and as national mandates for biofuels, especially in Europe, create new markets for vegetable oils that indirectly increase demand for palm oil.

But it's not just demand that is driving the expansion. Oil-palm plantations are a hot target for investors, whether from agribusiness, pension funds or corrupt tycoons looking for a safe and profitable way to launder funds. These days money is flowing into the bank accounts of palm oil companies, and they are using this cash to expand their land banks.

Producing all this cheap palm oil exacts a high price. Destruction of rainforests, labor exploitation, and brutal land grabbing: these are just a few of the nasty consequences that come with today's oil-palm plantations. And, with growing demand, those consequences are spreading out to more parts of the planet.

The global expansion of oil-palm plantations can only extend so far, however. Since oil-palms can only be cultivated economically in tropical areas near to the equator where there are high levels of rainfall, the global expansion of oil-palm plantations is concentrated in certain parts of Asia, Africa and Latin America where these conditions exist. It so happens that these lands are occupied by peasants and indigenous peoples and the tropical forests that they depend on.

The expansion of oil-palm plantations, therefore, is necessarily a story about the displacement of these people and the destruction of their forests and farms to make way for monoculture plantations.

Africa is at the center of the current push to expand oil-palm plantations. Oil-palms are not new to the continent. Africa is where the history of palm oil began. For generations, Africans have used oil-palms to produce vegetable oils, wines, medicines and numerous other products that are central to their economies, cultures and daily lives. Various attempts were made by the European powers to turn the crop into an industrial export commodity, produced on large plantations. But most of their efforts eventually crumbled and industrial production only took off in Malaysia and Indonesia, where African oil-palms were introduced at the turn of the 20th century. In Africa, palm oil remains mainly a product of small-scale production and processing, in the hands of millions of peasants, most of them women.

This is on the verge of changing drastically. With lands in Indonesia and Malaysia becoming more difficult for palm oil companies to acquire, attention is shifting to Africa as a new frontier for low-cost palm oil production for export. Over the past decade and a half, foreign companies have signed over 60 deals covering nearly four million hectares in central and western Africa for the development of oil-palm plantations. The land grabs are already generating violent conflicts in several African countries.

The situation in Africa is a reminder that this brutal expansion of oil-palm is not simply about land. It is about a larger struggle over food systems and models of development. Will African palm oil be produced by African peasants or transnational corporations? Will it be produced by peasants on mixed farms and semi-wild palm groves? Or will these peasants be displaced to make way for large-scale, industrial plantations? These questions have implications beyond Africa. If the continent becomes a new frontier for low-cost palm oil, exports from Africa will affect farmers growing vegetable oil crops in other countries, such as India and Mexico. There is therefore solidarity in the struggles of Cameroon peasants against oil-palm plantations and the struggles of coconut farmers in India against palm oil imports. Solidarity is also to be found with the peasants of the Aguan Valley in Honduras, who are fighting against big landlords to stop the violent takeover of their small oil-palm farms and co-operatives that serve local markets.

Cheap oil

Fifty years ago you would be hard pressed to find foods made with palm oil unless you were in Central or Western Africa where the crop originates from. Today it's hard to avoid it. Palm oil is everywhere, especially in processed foods. Studies suggest it is contained in about half of the packaged foods on supermarket shelves, whether you are shopping in Shanghai, Durban or

Santiago.[1] You'll also find it in most soaps, cosmetics and lotions.

The demand for palm oil is insatiable. Consumption has increased by about 1.5 million tonnes per year since the mid 1980s, going from just a few million tonnes to over 50 million tonnes today. Palm oil now accounts for over half of the world's total consumption of oils and fats.[2,3]

The underlying reason for the dramatic boom is simple: palm oil is cheap. Amongst the big crops for oils and fats (oil-palm, soybeans, oilseed rape and sunflower), palm oil is the cheapest.[4] So wherever there's a demand for a cheap, generic source of vegetable oil, palm oil tends to win out.[5]

This wasn't always the case. Not long ago, national markets for vegetable oil were dominated by local sources of oils and fats and national policies and regulations protected domestic vegetable oil producers from cheap imports. But over the past 15 years, the World Trade Organization (WTO) and a series of bilateral free trade agreements have removed most of these protections, opening the flood gates to vegetable oil imports.

Malaysian palm oil companies jumped on this opportunity. They expanded production, first in Malaysia and then in Indonesia. Other companies followed suit. Today, Malaysia's forests and agricultural lands are carpeted with oil-palm plantations, as are several islands of the Indonesian archipelago. These two palm oil powerhouses now account for around 90 per cent of global palm oil production and exports, a huge figure considering that palm oil accounts for nearly two-thirds of total global vegetable oil exports.

The surge in palm oil exports has hit farmers hard in importing countries, such as India. During the 1980s and early 1990s, the Indian government used import restrictions and government programs to maintain national self sufficiency in vegetable oil production. Decent prices encouraged farmers to expand into oilseeds and boost production of traditional vegetable oil crops, like coconut, whose production doubled in the first half of the 1990s. The local processing of the oils also generated thousands of jobs.

But in 1994, under pressure from the World Bank and as part of its WTO obligations, India started eliminating restrictions on vegetable oil imports. The country was immediately inundated with imported palm oil, while production of traditional oil crops languished. Today, with tariffs on palm oil fluctuating around zero, India is the world's largest importer and consumer of palm oil.[6]

China has followed a similar path. Imports of oil-palm hovered around one million tonnes per year until China made significant cuts to import restrictions in 2000 as part of its WTO entry obligations. Imports ballooned, reaching five million tonnes per year by 2005. That year, China began implementation of a free trade agreement with the Association of Southeast Asian Nations, which the Malaysian oil-palm industry credits with a further 34-per-cent increase in oil-palm imports between 2005 and 2010.[7]

The world's fourth-largest import market for palm oil, Pakistan, is also a product of free trade. The Malaysian palm oil industry says the 2008 Malaysia-Pakistan free trade agreement is responsible for doubling Pakistan's palm oil imports between 2007 and 2010.[8]

China, India and Pakistan were marginal consumers of palm oil two decades ago. Today they account for more than 40 per cent of total global imports, and a third of global consumption.[9]

Trade policies are not the only factor, however. The surge in palm oil imports in India and China, and in many other countries in the South, such as Venezuela and Bangladesh, also correlates with major transformations to their food systems. Global food corporations, restaurant chains and supermarkets are expanding rapidly in the South and this is increasing the consumption of processed foods. Annual sales growth of processed foods is around 29 per cent in low- and middle-income countries, as opposed to only seven per cent in high-income countries.[10]

More consumption of processed foods means more consumption of fats, and more consumption of palm oil, the world's discount source of fats.[11] In China, where supermarkets are expanding faster than anywhere else on earth, the annual per-capita consumption of vegetable oils has gone from three kilograms in 1980 to 23 kilograms in 2009, or roughly 64 grams per day – almost twice the fat intake required to meet a person's nutritional requirements. Palm oil now accounts for a third of the vegetable oil consumed in China, nearly three times the share it held in 1996.[12]

In Mexico, sales of processed foods have increased by 5 per cent to 10 per cent per year since the country began implementing the North American Free Trade Agreement with the US and Canada, opening the door to increased foreign investment by transnational food companies.[13] Obesity rates are soaring; Mexico now has a higher percentage of obese people than the US. And palm oil consumption is soaring too. Per-capita consumption of palm oil doubled from 1996, when it accounted for 11 per cent of the vegetable oil in the average Mexican's diet, to 2009, when it accounted for 28 per cent.[14]

Even in the US, there has been a recent shift by food companies towards the use of palm oil, partly in response to concerns over transfats. Since 2000, consumption of palm oil in the US has grown nearly sixfold.[15]

This is still well behind Europe, where consumption totalled 5.8 million tonnes in 2012, double what it was in 2000. Growth of palm oil consumption in Europe is, however, driven less by changes to the food system, as it is by the continent's biofuel policies. The implementation of biofuels mandates in European countries over the past decade or so has created much greater demand for palm oil, both as a feedstock for biodiesel and as a vegetable oil to replace European oilseeds that are diverted to biofuel production. Palm oil imports could surge much further if a European Commission proposal goes forward which would see all 27 members of the European Union requiring biofuels based on food crops to account for at least five per cent of national transport fuel consumption The legislation would require an additional 21 Mtoe (million tonnes oil equivalent) of biofuels by 2020. Measured in palm oil, this equates to roughly 5.5 million hectares of new oil-palm plantations.[16]

Cash crop

The surging global demand for oil-palm has produced windfall profits for palm oil companies and turned them into hot targets for investment by banks, pension funds and other financiers looking to cash in on the palm oil boom. All the major palm oil companies are plowing this new-found money into more plantations. So much so that it is difficult to say if money is a bigger driver of plantation expansion than the global demand for palm oil.

In Indonesia alone, $12.5 billion is estimated to have been invested in oil-palm expansion between 2000 and 2008, and those numbers are rising.[17] Much of this money comes from Singapore, where Indonesian tycoons parked their ill-got fortunes to escape from the clampdown that occurred after the collapse of the Suharto dictatorship. These tycoons are using this money and their old political influence to build up plantation empires throughout Indonesia, and even to other countries such as the Philippines and Liberia. Oil-palm plantations are also a favorite destination of Malaysian companies with close connections to the ruling elites. The forestry companies linked to the chief minister of the Malaysian state of Sarawak are particularly active in building up land banks for oil-palm plantations in Borneo, Papua and Africa.

Meanwhile in Colombia and Honduras, paramilitary groups and drug barons are deeply intertwined with oil-palm expansion.[18]

More conventional conduits for funnelling money into oil-palm plantations are also to be found. Some of the largest palm oil companies have recently turned to public offerings on stock markets to raise money from the financial houses and institutional investors eager to get a piece of the palm oil boom. In 2012, Felda, the Malaysian state palm oil company, restructured and went partially public, raising $3.3 billion in what was the third-largest share offering in the world that year. The share sale left Felda with a $2 billion cash pile that the company has since been using to acquire lands for oil-palm and rubber plantations outside of Malaysia.

That same year, one of the largest oil-palm plantation companies in Indonesia also made an initial public offering. Bumitama Agri, controlled by Indonesian billionaire Lim Hariyanto Wijaya Sarwono, raised around $177 million on the Singapore Stock Exchange, as palm oil giant Wilmar and several Asian, European and US financial management companies each bought multimillion dollar stakes in the company. Bumitama said it would allocate $114 million from the IPO for the expansion and development of its existing uncultivated land bank.[19]

A scramble for lands

With all this money pouring into palm oil companies, lands for oil-palm plantations are at an all time premium, wherever they can be found. Oil-palm plantations can, however, only be established on a narrow band of lands in tropical areas that are roughly seven degrees north or south of the equator and that have abundant and evenly spread rainfall. This makes the potential area for new oil-palm plantations rather limited. Plus, most of these lands are composed of forests and farmlands that are occupied by indigenous peoples and peasants, some of whom are already growing oil palms for local markets.

The expansion of oil-palm plantations, therefore, depends upon companies getting these people to give up their lands. This is not an easy sell, given the meagre jobs and other benefits that an oil-palm plantation generates in comparison with the destruction that it causes and the value that the lands already hold for the people. A typical oil-palm plantation requires only one poorly paid worker for every 2.3 hectares, while the surrounding communities pay a high price for the deforestation, water use, soil erosion and chemical fertilizer and pesticide contamination that it causes.[20]

The easy way for companies to get around these hurdles is to ensure that the communities do not even know that their lands have been signed away. It is very common in Africa, for instance, for companies to sign land deals directly with the national government without the knowledge of the affected communities. In many cases, the companies signing the deals are obscure companies registered in tax havens with their beneficial owners hidden from view. The managers of these companies tend to come from the mining sector or other extractive industries with long histories of shady deals in Africa. In Papua New Guinea and Indonesia, land deals are typically brokered between local elites and foreign investors, also often with obscure ownership structures registered in tax havens.

Such small shell companies are not in the business of developing plantations. Once the land contracts are signed, they immediately look to sell out to larger companies with the technical capacity and financial resources to build the plantations. And it is usually at this point that the communities come to understand that their lands have been sold.

Most of these cases eventually lead to a situation where a large transnational plantation company, backed by a national government and a multi-million dollar contract, faces off against a poor community trying desperately to hold on to the lands and forests it needs to survive. It is incredibly difficult for communities to defend themselves against such powerful forces, and those that do risk the treat of violence, whether by paramilitaries in Colombia, police in Sierra Leone or the army in Indonesia.

Communities lose out from oil-palm plantations

Local communities can only lose from this new wave of land grabs for palm oil. They lose access to vital lands and water resources, now and for future generations. And they have to face all of the impacts that come with vast monoculture plantations within their territories – pollution from pesticides, soil erosion, deforestation, and labor migration. Experience also shows that the employment generated by the plantations often goes to outsiders, and that most of the jobs are seasonal, poorly paid, and dangerous. Certification schemes, such as the Roundtable on Sustainable Palm Oil (RSPO), can only alleviate or postpone some of the worst excesses.

Experience also shows that the outgrower schemes, known as plasma programs in Indonesia or nucleus estates in Africa, are not solutions. It has become standard practice for companies to offer to develop outgrower schemes on a portion of the lands within their concessions as part of their agreements with host governments. The farmers involved in these schemes have little control over production or the terms of payment, which are dictated by the company, leaving them vulnerable to all sorts of abuses. More than anything, the outgrower schemes are a means for the companies to capture supply and placate the local people who are ultimately being forced to give up control over their territories.

This is not to say that small-scale palm oil production cannot support people's livelihoods. There are excellent examples from Honduras and West and Central Africa where small oil-palm farmers have developed markets or organized co-operatives that provide them with a decent price for their production.[21] But in these cases, the farmers have control over their lands and their farms, and they are not at the mercy of a single foreign or national company for the sale of their products. The current wave of plantations is a direct threat to these farmers – taking away their lands and their local markets.

There is no demand justification for the expansion of oil-palm plantations either. The growing global market for palm oil is not about resolving world hunger. It is mainly a product of new biofuel mandates and the substitution of cheap imported palm oil for locally produced oils and fats (whether animal or vegetable) in the production of processed foods by global corporations. People do not need more oil-palm plantations; corporations do.

Africa: another side of palm oil

There is a part of the world where palm oil is not synonymous with defor-estation and plantations, where it is not an export commodity but an essential ingredient in local dishes, and where its production profits peasants not bankers. In Africa, the center of origin for oil-palm, tens of millions of people, most of them women, rely on this tree for food and livelihoods.

The global land grab for plantations puts these people, the oil-palms they look after and their traditional systems of production at tremendous risk. Resistance, for them, is not just a matter of holding on to their lands and forests. It is also a fight for their livelihoods, their cultures, their biodiversity and their food sovereignty.

From the plantations of Malaysia to the small farms of Honduras, all oil-palms trace their origins to Africa. It was here, long ago, somewhere in the western and central parts of the continent that people first began to use the plant for their needs. They discovered dozens of uses for the plant, and it soon became an integral part of their food systems and local economies and cultures. In the traditional songs of many countries of West and Central Africa, oil-palm is called the "tree of life".[22]

In Africa, oil-palms on plantations make up only a small percentage of the total. Most oil-palms are still grown in the groves in mixed forests. These groves are often cared for and harvested by particular families, passed down from generation to generation. Such semi-wild groves are found in large parts of Africa. Nigeria contains the continent's largest area of wild or semi-wild palm groves, with over 2.5 million hectares. Oil-palms are also grown on small farms on the continent. African farmers in West and Central Africa mix oil-palms with other crops like bananas, cacao, coffee, groundnuts and cucumbers.

In the local markets of West and Central Africa, the quality of a palm oil is typically judged by its color. African women say that the palm oil extracted from traditional oil-palms is better because it is redder than that extracted from the modern varieties. In Benin, traditional palm oil sells for between 20 and 40 per cent more in the markets than that from modern varieties.[23] African women also say that their traditional sauces made with boiled palm kernels have a lighter and thus better texture when made with kernels from traditional palms than with those from modern ones. The economic importance of oil-palms to Africa is huge, particularly when it comes to women. They handle most of the production, from the harvest and processing of palm oil, to the sale of the oil and other oil-palm products in the local markets. The income they earn makes a critical contribution to their households. In the south of Benin, for example, around one quarter of all women earn some part of their income from the processing and sale of palm oil.[24]

But traditional oil-palms offer much more than a high-quality palm oil and kernel. Unlike on industrial oil-palm plantations, African communities use every part of a traditional oil-palm, from its roots to its branches, to produce everything from wines and soups to soaps and ointments, traditional medicines and animal feeds, and even a whole range of textiles and housing materials

The global land grab for oil-palm plantations hits Africa

The rush to develop oil-palm plantations in Africa is a double-whammy for the continent. Not only does it involve a huge land grab of people's lands and food producing resources, it also directly undercuts the livelihoods of millions of people involved in Africa's traditional oil-palm sector.

This is not the first time foreigners have pushed an expansion of oil-palm in Africa. During the colonial occupation of the continent, the European powers became interested in palm oil as an industrial lubricant and for making candles. African families were forced to pay a special tax, known as the "takouè" to the colonial authorities, in the form of palm oil and palm nut. King Léopold II of Belgium forced every farmer in the province of Equateur in the Congo to plant 10 palms a year.[25]

With independence, most of these plantations and research stations were nationalized, and the new African governments re-energized the expansion of national production. But, at the end of the 1990s, World Bank and donor-imposed structural adjustment programs forced African governments to privatize their national palm oil companies and to sell off their mills and plantations. While many national companies simply crumbled away, European companies with old colonial connections captured the most lucrative operations.

Today there is a second wave of foreign interest in oil-palm plantations in Africa. With land for oil-palm plantations becoming more difficult and expensive to acquire in Malaysia and Indonesia, companies and speculative investors are keen to open up new frontiers for export production. Some investment is going to Papua and to Latin America, but the biggest target is Africa. A long list of companies, from Asian palm oil giants to Wall Street financial houses, are scrambling to get control over lands on the continent that are favorable to oil-palm, especially in the West and Central regions.

Resistance is building

The communities facing land grabs from palm oil companies are under tremendous pressure to accommodate them, with pressure coming from the companies, the government, the local chiefs and even the army and para-militaries. Those who resist face arrest, harassment and violence. And yet communities in Africa and around the world, from Papua New Guinea to Sarawak, from Cameroon to Guatemala, continue to struggle to stop palm oil companies from entering their lands.

Controlling the food system

Communities in southwest Cameroon have been involved in a three-year struggle to stop the US company Herakles Capital from setting up an oil-palm plantation in their area. Despite support from the president of Cameroon, Herakles has been unable to move forward with its plans because the communities are united in their total opposition to the plantation and because of the creative actions that they have undertaken, with support from national and international partners, to put pressure on the company to leave. The company and the government keep coming back and presenting new terms, the latest being a presidential decree that reduces the land allocated to Herakles from 73,000 hectares to 20,000 hectares and boosts the rent that the company must pay. Community leaders have been arrested and harassed with lawsuits. Yet the communities are sticking to their bottom line demand – no oil-palm plantations on their lands.

Cameroon is also a target for the Luxembourg-based company SOCFIN, owned by billionaires Vincent Bolloré of France and Hubert Fabri of Belgium. Over the past decade and a half, SOCFIN has taken over lands for oil-palm and other crops in several African countries, including Cameroon, DRC, Guinea, Nigeria, Sao Tome and Principe, and Sierra Leone. The company is notorious for human rights abuses and land conflicts at its operations, and for its aggressive tactics against those who oppose it. In the past few years, the company has slapped defamation suits on several organizations and journalists in Africa and Europe that have spoken out against it.

On 5 June 2013, communities affected by SOCFIN plantations in four African countries held simultaneous protest actions against the company, as a delegation of diaspora from these countries and supported by the French group Réseaux d'Action Transnationale (ReAct) presented a joint letter from the various communities to the Annual General Meeting of the Bolloré Group, which is a major shareholder in SOCFIN. "This initial international protest is just the beginning. We are committed to upholding our rights and Mr Bolloré will have to understand that," said Emmanuel Elong, spokesperson of Synaparcam, the Socapalm resident farmers' union in Cameroon.[26]

Strong community resistance combined with national and international well-targeted pressure, can roll back land grabs. The Jogbahn Clan in Liberia provides an inspiring example. When the British company Equatorial Palm Oil began surveying their lands as part of a deal it signed with the Liberian government, the communities took action to stop the work crews. They then marched to the local government offices to make it clear that they had never been consulted about the deal and that they would never give up their lands for the project. Along the way they were beaten, arrested and thrown in jail. But the communities refused to back down. Local and international NGOs joined their struggle, and exposed what was happening to the world. Finally, in March 2014, community leaders met with the Liberian President, Ellen Johnson Sirleaf, and secured a commitment from her to stop the company from expanding on their lands. Now Liberian groups are hoping to replicate these efforts with other affected communities in the country.[27]

The many different efforts to resist land grabs and maintain local control over palm oil production in Africa, Asia and Latin America demonstrate how committed local communities are to maintaining control over their ancestral lands and their biodiversity, for themselves and for future generations.

This article is extracted from a GRAIN booklet. Full text available here: https://www.grain. org/e/5031

1   RSPO (2013) "Why palm oil matters in your everyday life", http://www.rspo.org/file/VISUAL%20-%20Consumer%20Fact%20Sheet.pdf

2   IEA Bioenergy (2009) "A global overview of vegetable oils, with reference to biodiesel", June, http://www.bioenergytrade.org/downloads/vegetableoilstudyfinaljune18.pdf

3   Sime Darby Plantation (2014) "Palm oil facts & figures", http://www.simedarby.com/upload/Palm_Oil_Facts_and_Figures.pdf

4   *Ibid.*

5   GRAIN thanks Pastor Adjahossou Firmin, resource person on palm oil from Benin, for his contributions.

6   Jafri, A. (2011) "Trade Liberalisation's Impact on Edible Oil Sector in India", Focus on the Global South, 6 July, http://siccfm.blogspot.ca/2012/01/trade-liberalization-and-impact-on.html

7   Balu, N. and Ismail, N. "Free Trade Agreement – The Way Forward for the Malaysian Palm Oil Industry", http://palmoilis.mpob.gov.my/publications/OPIEJ/opiejv11n2-balu.pdf (11.2)

8   *Ibid.*

9   Lipid Library, http://lipidlibrary.aocs.org/market/palmoil.htm

10 Hawkes, C. (2007) "Globalization and the Nutrition Transition: A Case Study", in: Per Pinstrup-Andersen and Fuzhi Cheng (editors), "Food Policy for Developing Countries: Case Studies", http://cip.cornell.edu/dns.gfs/1200428200

11 Rainforest Action Network (2013) "Conflict palm oil", September, http://www.ran.org/conflict_palm_oil

12 Sharma, D.C. (2012) "Rise in oil consumption by Indians sets off alarm", *India Today*, 2 April, http://indiatoday.intoday.in/story/rise-in-oil-consumption-by-indians-sets-off-alarm/1/182679.html

13 Hawkes, C. (2007) *op.cit.*

14 FAOSTAT
15 Rainforest Action Network (2013) *op.cit.*
16 GRAIN (2013) "Land grabbing for biofuels must stop", 21 February, https://www.grain.org/article/entries/4653-land-grabbing-for-biofuels-must-stop
17 Pacheco, P. (2012) "Oil palm in Indonesia linked to trade and investment", CIFOR, May, http://www.cifor.org/ard/documents/results/Day2_Pablo%20Pacheco.pdf
18 See the sections on regional expansion of oil palm plantations
19 A 2013 Friends of the Earth report shows how Bumitama's land bank is composed of thousands of hectares of lands that the company is operating on illegally, without the necessary permits, https://www.foeeurope.org/sites/default/files/press_releases/commodity_crimes_nov13.pdf
20 UNEP Global Environmental Alert Service (2011) "Oil palm plantations: threats and opportunities for tropical ecosystems", December, http://www.unep.org/pdf/Dec_11_Palm_Plantations.pdf
21 For examples from Honduras, see Kerssen, T.M. (2010) *Grabbing Power: The New Struggles for Land, Food and Democracy in Northern Honduras*, FoodFirst Books; for examples from Africa, see World Rainforest Movement (2013) *Oil palm in Africa: past, present and future scenarios*, WRM
22 Based in part on field studies that were carried out by GRAIN in 2013 in collaboration with Réseau des Acteurs du Développement Durable (RADD) in Cameroon, INADES – Formation (Institut Africain pour le Développement Economique et Social) in Côte d'Ivoire, ONG ADAPE (Association pour le Développement Durable et la Protection de l'Environnement) in Guinea and CEPECO (Centre pour la Promotion et l'Education des Communautés de base) in the Congo.
23 Fournier, S. et al (2001) "Importance of 'local' commodities: palm oil production in Benin", *Oléagineux, Corps Gras, Lipides*, 8(6): 646–53, November–December
24 *Ibid.*
25 World Rainforest Movement (2010) "Oil palm in Africa: past, present and future scenarios"
26 Synaparcam, SoGB residents committee, Concern Union Citizen, and MALOA (2013) "West African farmers stand up against Bolloré", 5 June, http://farmlandgrab.org/post/view/22157
27 For more information about the case see: http://sdiliberia.org/node/263

4.6 Free trade and Mexico's junk food epidemic

Transnational food companies understand that their main "growth markets" are now in the Global South. To increase their profits they need to develop and sell products aimed at hundreds of millions of the world's poor. Many of these people and communities still eat food that they produce themselves, or buy from informal markets that sell local produce. For many, these local food systems and circuits are also their livelihood.

To reach these potential consumers, large food corporations are infiltrating, inundating and taking over traditional distribution channels and replacing local foods with cheap, highly processed industrial foods, often with the active support of the Mexican governments (at national, municipal and even local level).

Many people, in both rural and urban areas, are trapped in a situation where the only available food is of poor quality, heavily processed or just plain junk food. Free trade and investment agreements have been critical to the success of transnational food corporations. The case of Mexico provides a stark illustration.

Over the past two decades, the Mexican government has signed more than a dozen free trade agreements and nearly 30 investment treaties, which have opened up the countryside and the retail sector to transnational companies, putting Mexico's food system up for grabs. The North American Free Trade Agreement, signed in 1993, triggered an immediate surge of direct investment from the US into Mexico's food processing industry. Between 1999 and 2004, three-quarters of the country's foreign investment went into the production of processed foods. At the same time, sales of processed foods went up by 5 per cent to 10 per cent each year.

Mexico is now one of the 10 biggest producers of processed food in the world, with total sales reaching $124 billion in 2012. The corporations involved – including PepsiCo, Nestlé, Unilever and Danone – made $28 billion in profits from these sales, $9 billion more than they made in Brazil, Latin America's largest economy.

Mexico offers the food processing industry not only low costs (a saving of 14.1 per cent compared to the US), but also, according to Roberto Morales, writing in *El Economista*, a "network of trade agreements that permit access to big markets such as the European Union and the US with tariff preferences". Even with the global economic crisis, "the sales of the retail business establishments have grown steadily in the last three years". These corporations are investing heavily in taking over local distribution. Huge supermarkets are an important part of these emerging patterns of retail distribution because they concentrate goods, but the crucial aim is to replace the corner shops, thus aggressively taking over former independent territories of trade.

The food corporations began by colonizing the existing, dominant food distribution networks of small-scale vendors, the *tiendas* (the corner shops). There are some 400,000 *tiendas*, *estanquillos* or *misceláneas* premises in Mexico, stores smaller than 10 square meters, which carry a limited variety of products and are equipped with a limited amount of refrigeration and inventory.

Food corporations swamped the *tiendas*' distribution channels and lowered transportation costs for their own products (for example, PepsiCo links the delivery of various goods that it produces to each destination, thus undercutting rivals that don't have this practice). This kind of move boosted enormous sales because of what the industry calls "the absolute domination of the sales point": a drastic reduction in the options to buy. People will eat what they have at hand, and suddenly the only available goods were a narrow range of packaged, bottled or canned goods.

In many neighborhoods, and even in the countryside's isolated communities, this kind of processed food became the only stuff available. Controlling availability became a crucial factor in the processed food business and when convenience stores (some of them owned by the big processed food companies) displaced tiendas, this control grew. According to the Mexican Chamber of Commerce, five *tiendas* close for every convenience store that opens.

By 2012, retail chains had displaced *tiendas* as Mexico's main source of food sales, controlling 35 per cent of the country's market; *tiendas* held on to 30 per cent and open street markets to 25 per cent. The remaining 10 per cent, sometimes not accounted for, is held by hotels, restaurants and cafés.

The number of supermarkets, discount chains and convenience stores exploded from 700 to 3,850 in 1997 alone, and by 2004 numbered 5,730. Today, Oxxo, a convenience store chain owned by Femsa, a unit of Coca-Cola Mexico, is opening an average of three stores a day, and aims to inaugurate its 14,000th store this year.

This boom is perfectly in line with a territorial control aimed at disabling and wiping out the corner stores, fighting street by street to impose the corporate vision of food consumption. Big supermarkets will persist, but it is convenience stores that will let corporations reach the poorest populations in their own neighborhood.

With this move onto every street corner, transnational food companies that produce and sell processed food (and privilege these over fresh food), are set to dominate the scene in terms of production, territory and sales. This represents the ultimate control of the sale's point, an almost total control over availability.

One of the main effects of all this has been a radical change in people's diets and a disproportionate increase in malnutrition, obesity and diabetes. Mexico's National Institute for Public Health reports that between 1988 and 2012, the proportion of overweight women between the ages of 20 and 49 increased from 25 per cent to 35.5 per cent and the number of obese women in this age group increased from 9.5 per cent to 37.5 per cent. A staggering 29 per cent of Mexican children between the ages of 5 and 11 were found to be overweight, as were 35 per cent of the youngsters between 11 and 19, while 1 in 10 school-age children suffers from anemia.

The level of diabetes is equally troubling. The Mexican Diabetes Federation says there are up to 10 million people who suffer from diabetes in Mexico, with about two million of them unaware that they have the disease. This means that more than seven per cent of the Mexican population has diabetes. The incidence rises to 21 per cent for people between the ages of 65 and 74. Diabetes is now the third most common cause of death in Mexico, directly or indirectly. In 2012, Mexico ranked sixth in the world for diabetes deaths. Specialists predict that there will be 11.9 million Mexicans with diabetes by 2025.

Obesity and diabetes function together, interacting so strongly that a new term has emerged: "diabesity". Who to thank for this? The transnational food industry, supported by governments that share their interests.

After visiting the country in 2012, the UN Special Rapporteur on the Right to Food got it right when he said, "The trade policies currently in place favor greater reliance on heavily processed and refined foods with a long shelf life rather than on the consumption of fresh and more perishable foods, particularly fruit and vegetables. The overweight and obesity emergency that Mexico is facing could have been avoided, or largely mitigated, if the health concerns linked to shifting diets had been integrated into the design of those policies."

Alongside this invasion of processed food linked to the aggressive takeover of the corporative mini-marts, Mexico's public policies on food, including a national "crusade against hunger", are closely associated with the big food corporations (including PepsiCo and Nestlé). In Mexico, their influence is so great that some government assistance programs officially promote their products.

Big farms and producers are well looked after, while peasants and small-scale producers get almost no support. According to the former Special Rapporteur on Food, "less than 8 per cent of expenditures on agricultural programs" benefit this poor sector.

Unfortunately, Mexico is in the hands of the big corporations.

To read more about this issue, please see GRAIN's longer and fully referenced report,
"Free trade and Mexico's junk food epidemic" at https://www.grain.org/e/5170